The Gateway tests are end-of-course tests in English which Tennessee students must pass as part of the r₀ , high school diploma.

These tests will count fifteen percent (15%) of the grade in the course for all students, including Special Education students.

Conditions that most schools will enforce during the Gateway test:

- The Algebra teacher cannot be in the room when the students are being tested.
- The State discourages large group testing.
- **There is no time limit on the Gateway Tests.**
- Calculators: Math teachers need to tell students before the test is given that the following calculators are prohibited: TI89, TI92;Casio CFX 9970G; Casio Algebra FX 2.0, HP 49G; and any pocket organizer. If a student does not have a calculator, one should be provided.

Students will be given additional opportunities throughout their high school career -- during the school year and in the summer -- to retake any of the Gateway tests failed.

This workbook and video are designed to assist students in successfully passing the Algebra I Gateway Exam. The video explains strategies for solving problems for those students who struggle with algebra concepts.

Most of the test questions have more than one way of being solved. In many cases a student can eliminate one or more answers even before working out the problem.

After completing the practice tests and watching the video, a student should be able to identify the types of problems where answers can be eliminated. The student should also be able to use different strategies for finding the correct answer to a question.

To use this program, start by taking practice test 1. After completing the test, check your answers with the answer key. Then watch the video of explanations for test one.

Next, take practice test two. After completing the test check your answers with the answer key. Then watch the video of explanations for test two.

Finally, try taking test three. The explanations for test three are written at the end of the workbook.

It would be best to rest for a couple of days between taking the practice tests.

Be sure to get a good nights sleep the day before the test. Eat a good breakfast the morning of the test.

When taking the test, do not skip problems thinking you will go back to work on it later. Many students finish the test and forget to go back. Also, answer every question. Answers left blank will be counted as wrong.

Remember, there is no time limit on the test. Even though there is no time limit, do not waste a lot of time and energy on a single problem. This may tire you out and cause you to make mistakes on easier problems.

When you get to a difficult problem; guess, write down the number on your scratch sheet of paper, and move on. When you finish the other problems you can go back. If you forget to go back, at least you have filled in an answer.

When taking the test: use a familiar calculator. Do not borrow one the day of the test unless you have no other choice.

Read the directions carefully. If you do not understand the directions, you may ask for an explanation.

You may find more sample questions at www.state.tn.us/education/ .

Good luck.

Algebra I Reference Page
(Students will be given a reference page to use during the exam.)

d = rt $\qquad$ distance = rate × time

Distance Formula: $\qquad d = \sqrt{(x_2 - x_1) + (y_2 - y_1)}$

Point-Slope Equation: $\qquad y - y_1 = m(x - x_1)$

Pythagorean Theorem: $\qquad a^2 + b^2 = c^2$

Slope Formula: $\qquad m = \dfrac{y_2 - y_1}{x_2 - x_1}$

Slope-Intercept Equation:` $\qquad$ y = mx + b

$\pi = 3.14$ or $\dfrac{22}{7}$

n	$\sqrt{n}$	n^2
1	1.000	1
2	1.414	4
3	1.732	9
4	2.000	16
5	2.236	25
6	2.449	36
7	2.646	49
8	2.828	64
9	3.000	81
10	3.162	100
11	3.317	121
12	3.464	144
13	3.606	169
14	3.742	196
15	3.873	225
16	4.000	256
17	4.123	289
18	4.243	324
19	4.359	361
20	4.472	400
21	4.583	441
22	4.690	484
23	4.796	529
24	4.899	576
25	5.000	625

PERIMETER (P) AND CIRCUMFERENCE (C)

Any Polygon: $\qquad$ P = sum of side lengths
Rectangle: $\qquad$ $P = 2l + 2w$
Circle: $\qquad$ $C = 2\pi r$ or πd

PLANE FIGURES $\qquad\qquad$ AREA (A)

Triangle $\qquad A = \dfrac{1}{2}bh$

Rectangle $\qquad A = lw$

Circle $\qquad A = \pi r^2$

SOLID FIGURES $\qquad\qquad$ VOLUME (V)

Prism $\qquad$ V = Bh
$\qquad\qquad\qquad\qquad$ or
$\qquad\qquad\qquad\qquad$ V = lwh

Cube $\qquad$ V = s^3

ABBREVIATIONS
A = area
B = area of base
b = base
C = circumference
d = diameter
h = height
l = length
P = perimeter
r = radius
s = length of side
V = volume
w = width

Practice

Test

1

1. Ⓐ Ⓑ Ⓒ Ⓓ
2. Ⓕ Ⓖ Ⓗ Ⓙ
3. Ⓐ Ⓑ Ⓒ Ⓓ
4. Ⓕ Ⓖ Ⓗ Ⓙ
5. Ⓐ Ⓑ Ⓒ Ⓓ
6. Ⓕ Ⓖ Ⓗ Ⓙ
7. Ⓐ Ⓑ Ⓒ Ⓓ
8. Ⓕ Ⓖ Ⓗ Ⓙ
9. Ⓐ Ⓑ Ⓒ Ⓓ
10. Ⓕ Ⓖ Ⓗ Ⓙ
11. Ⓐ Ⓑ Ⓒ Ⓓ
12. Ⓕ Ⓖ Ⓗ Ⓙ
13. Ⓐ Ⓑ Ⓒ Ⓓ

14. Ⓕ Ⓖ Ⓗ Ⓙ
15. Ⓐ Ⓑ Ⓒ Ⓓ
16. Ⓕ Ⓖ Ⓗ Ⓙ
17. Ⓐ Ⓑ Ⓒ Ⓓ
18. Ⓕ Ⓖ Ⓗ Ⓙ
19. Ⓐ Ⓑ Ⓒ Ⓓ
20. Ⓕ Ⓖ Ⓗ Ⓙ
21. Ⓐ Ⓑ Ⓒ Ⓓ
22. Ⓕ Ⓖ Ⓗ Ⓙ
23. Ⓐ Ⓑ Ⓒ Ⓓ
24. Ⓕ Ⓖ Ⓗ Ⓙ
25. Ⓐ Ⓑ Ⓒ Ⓓ
26. Ⓕ Ⓖ Ⓗ Ⓙ

27. Ⓐ Ⓑ Ⓒ Ⓓ
28. Ⓕ Ⓖ Ⓗ Ⓙ
29. Ⓐ Ⓑ Ⓒ Ⓓ
30. Ⓕ Ⓖ Ⓗ Ⓙ
31. Ⓐ Ⓑ Ⓒ Ⓓ
32. Ⓕ Ⓖ Ⓗ Ⓙ
33. Ⓐ Ⓑ Ⓒ Ⓓ
34. Ⓕ Ⓖ Ⓗ Ⓙ
35. Ⓐ Ⓑ Ⓒ Ⓓ
36. Ⓕ Ⓖ Ⓗ Ⓙ
37. Ⓐ Ⓑ Ⓒ Ⓓ
38. Ⓕ Ⓖ Ⓗ Ⓙ
39. Ⓐ Ⓑ Ⓒ Ⓓ

40. Ⓕ Ⓖ Ⓗ Ⓙ
41. Ⓐ Ⓑ Ⓒ Ⓓ
42. Ⓕ Ⓖ Ⓗ Ⓙ
43. Ⓐ Ⓑ Ⓒ Ⓓ
44. Ⓕ Ⓖ Ⓗ Ⓙ
45. Ⓐ Ⓑ Ⓒ Ⓓ
46. Ⓕ Ⓖ Ⓗ Ⓙ
47. Ⓐ Ⓑ Ⓒ Ⓓ
48. Ⓕ Ⓖ Ⓗ Ⓙ
49. Ⓐ Ⓑ Ⓒ Ⓓ
50. Ⓕ Ⓖ Ⓗ Ⓙ
51. Ⓐ Ⓑ Ⓒ Ⓓ
52. Ⓕ Ⓖ Ⓗ Ⓙ

53. Ⓐ Ⓑ Ⓒ Ⓓ
54. Ⓕ Ⓖ Ⓗ Ⓙ
55. Ⓐ Ⓑ Ⓒ Ⓓ
56. Ⓕ Ⓖ Ⓗ Ⓙ
57. Ⓐ Ⓑ Ⓒ Ⓓ
58. Ⓕ Ⓖ Ⓗ Ⓙ
59. Ⓐ Ⓑ Ⓒ Ⓓ
60. Ⓕ Ⓖ Ⓗ Ⓙ
61. Ⓐ Ⓑ Ⓒ Ⓓ
62. Ⓕ Ⓖ Ⓗ Ⓙ

Practice

Test

2

1. Ⓐ Ⓑ Ⓒ Ⓓ
2. Ⓕ Ⓖ Ⓗ Ⓙ
3. Ⓐ Ⓑ Ⓒ Ⓓ
4. Ⓕ Ⓖ Ⓗ Ⓙ
5. Ⓐ Ⓑ Ⓒ Ⓓ
6. Ⓕ Ⓖ Ⓗ Ⓙ
7. Ⓐ Ⓑ Ⓒ Ⓓ
8. Ⓕ Ⓖ Ⓗ Ⓙ
9. Ⓐ Ⓑ Ⓒ Ⓓ
10. Ⓕ Ⓖ Ⓗ Ⓙ
11. Ⓐ Ⓑ Ⓒ Ⓓ
12. Ⓕ Ⓖ Ⓗ Ⓙ
13. Ⓐ Ⓑ Ⓒ Ⓓ

14. Ⓕ Ⓖ Ⓗ Ⓙ
15. Ⓐ Ⓑ Ⓒ Ⓓ
16. Ⓕ Ⓖ Ⓗ Ⓙ
17. Ⓐ Ⓑ Ⓒ Ⓓ
18. Ⓕ Ⓖ Ⓗ Ⓙ
19. Ⓐ Ⓑ Ⓒ Ⓓ
20. Ⓕ Ⓖ Ⓗ Ⓙ
21. Ⓐ Ⓑ Ⓒ Ⓓ
22. Ⓕ Ⓖ Ⓗ Ⓙ
23. Ⓐ Ⓑ Ⓒ Ⓓ
24. Ⓕ Ⓖ Ⓗ Ⓙ
25. Ⓐ Ⓑ Ⓒ Ⓓ
26. Ⓕ Ⓖ Ⓗ Ⓙ

27. Ⓐ Ⓑ Ⓒ Ⓓ
28. Ⓕ Ⓖ Ⓗ Ⓙ
29. Ⓐ Ⓑ Ⓒ Ⓓ
30. Ⓕ Ⓖ Ⓗ Ⓙ
31. Ⓐ Ⓑ Ⓒ Ⓓ
32. Ⓕ Ⓖ Ⓗ Ⓙ
33. Ⓐ Ⓑ Ⓒ Ⓓ
34. Ⓕ Ⓖ Ⓗ Ⓙ
35. Ⓐ Ⓑ Ⓒ Ⓓ
36. Ⓕ Ⓖ Ⓗ Ⓙ
37. Ⓐ Ⓑ Ⓒ Ⓓ
38. Ⓕ Ⓖ Ⓗ Ⓙ
39. Ⓐ Ⓑ Ⓒ Ⓓ

40. Ⓕ Ⓖ Ⓗ Ⓙ
41. Ⓐ Ⓑ Ⓒ Ⓓ
42. Ⓕ Ⓖ Ⓗ Ⓙ
43. Ⓐ Ⓑ Ⓒ Ⓓ
44. Ⓕ Ⓖ Ⓗ Ⓙ
45. Ⓐ Ⓑ Ⓒ Ⓓ
46. Ⓕ Ⓖ Ⓗ Ⓙ
47. Ⓐ Ⓑ Ⓒ Ⓓ
48. Ⓕ Ⓖ Ⓗ Ⓙ
49. Ⓐ Ⓑ Ⓒ Ⓓ
50. Ⓕ Ⓖ Ⓗ Ⓙ
51. Ⓐ Ⓑ Ⓒ Ⓓ
52. Ⓕ Ⓖ Ⓗ Ⓙ

53. Ⓐ Ⓑ Ⓒ Ⓓ
54. Ⓕ Ⓖ Ⓗ Ⓙ
55. Ⓐ Ⓑ Ⓒ Ⓓ
56. Ⓕ Ⓖ Ⓗ Ⓙ
57. Ⓐ Ⓑ Ⓒ Ⓓ
58. Ⓕ Ⓖ Ⓗ Ⓙ
59. Ⓐ Ⓑ Ⓒ Ⓓ
60. Ⓕ Ⓖ Ⓗ Ⓙ
61. Ⓐ Ⓑ Ⓒ Ⓓ
62. Ⓕ Ⓖ Ⓗ Ⓙ

Practice
Test
3

1. Ⓐ Ⓑ Ⓒ Ⓓ
2. Ⓕ Ⓖ Ⓗ Ⓙ
3. Ⓐ Ⓑ Ⓒ Ⓓ
4. Ⓕ Ⓖ Ⓗ Ⓙ
5. Ⓐ Ⓑ Ⓒ Ⓓ
6. Ⓕ Ⓖ Ⓗ Ⓙ
7. Ⓐ Ⓑ Ⓒ Ⓓ
8. Ⓕ Ⓖ Ⓗ Ⓙ
9. Ⓐ Ⓑ Ⓒ Ⓓ
10. Ⓕ Ⓖ Ⓗ Ⓙ
11. Ⓐ Ⓑ Ⓒ Ⓓ
12. Ⓕ Ⓖ Ⓗ Ⓙ
13. Ⓐ Ⓑ Ⓒ Ⓓ

14. Ⓕ Ⓖ Ⓗ Ⓙ
15. Ⓐ Ⓑ Ⓒ Ⓓ
16. Ⓕ Ⓖ Ⓗ Ⓙ
17. Ⓐ Ⓑ Ⓒ Ⓓ
18. Ⓕ Ⓖ Ⓗ Ⓙ
19. Ⓐ Ⓑ Ⓒ Ⓓ
20. Ⓕ Ⓖ Ⓗ Ⓙ
21. Ⓐ Ⓑ Ⓒ Ⓓ
22. Ⓕ Ⓖ Ⓗ Ⓙ
23. Ⓐ Ⓑ Ⓒ Ⓓ
24. Ⓕ Ⓖ Ⓗ Ⓙ
25. Ⓐ Ⓑ Ⓒ Ⓓ
26. Ⓕ Ⓖ Ⓗ Ⓙ

27. Ⓐ Ⓑ Ⓒ Ⓓ
28. Ⓕ Ⓖ Ⓗ Ⓙ
29. Ⓐ Ⓑ Ⓒ Ⓓ
30. Ⓕ Ⓖ Ⓗ Ⓙ
31. Ⓐ Ⓑ Ⓒ Ⓓ
32. Ⓕ Ⓖ Ⓗ Ⓙ
33. Ⓐ Ⓑ Ⓒ Ⓓ
34. Ⓕ Ⓖ Ⓗ Ⓙ
35. Ⓐ Ⓑ Ⓒ Ⓓ
36. Ⓕ Ⓖ Ⓗ Ⓙ
37. Ⓐ Ⓑ Ⓒ Ⓓ
38. Ⓕ Ⓖ Ⓗ Ⓙ
39. Ⓐ Ⓑ Ⓒ Ⓓ

40. Ⓕ Ⓖ Ⓗ Ⓙ
41. Ⓐ Ⓑ Ⓒ Ⓓ
42. Ⓕ Ⓖ Ⓗ Ⓙ
43. Ⓐ Ⓑ Ⓒ Ⓓ
44. Ⓕ Ⓖ Ⓗ Ⓙ
45. Ⓐ Ⓑ Ⓒ Ⓓ
46. Ⓕ Ⓖ Ⓗ Ⓙ
47. Ⓐ Ⓑ Ⓒ Ⓓ
48. Ⓕ Ⓖ Ⓗ Ⓙ
49. Ⓐ Ⓑ Ⓒ Ⓓ
50. Ⓕ Ⓖ Ⓗ Ⓙ
51. Ⓐ Ⓑ Ⓒ Ⓓ
52. Ⓕ Ⓖ Ⓗ Ⓙ

53. Ⓐ Ⓑ Ⓒ Ⓓ
54. Ⓕ Ⓖ Ⓗ Ⓙ
55. Ⓐ Ⓑ Ⓒ Ⓓ
56. Ⓕ Ⓖ Ⓗ Ⓙ
57. Ⓐ Ⓑ Ⓒ Ⓓ
58. Ⓕ Ⓖ Ⓗ Ⓙ
59. Ⓐ Ⓑ Ⓒ Ⓓ
60. Ⓕ Ⓖ Ⓗ Ⓙ
61. Ⓐ Ⓑ Ⓒ Ⓓ
62. Ⓕ Ⓖ Ⓗ Ⓙ

Gateway Mathematics Practice Test 1

1. 1. Simplify $x \cdot x \cdot y \cdot x \cdot y \cdot y \cdot x$
   A. $4x + 3y$
   B. $x^4 + y^3$
   C. $x^4 y^3$
   D. $12xy$

2. Find the next number in the sequence.

   2, 7, 12, 17,___

   F. 24
   G. 25
   H. 20
   J. 22

3. Which of the following sets of numbers is listed from least to greatest?

   A. 2.4, 3.1, 4, -5.3, -6.9, -7.1
   B. -2, 3, -4, 5, -6
   C. -3.1, -2, 1, 3.4, 5.2
   D. 3.5, 2.9, 0, -1.1, -2.3

4. On an English writing exam 15 students wrote with a blue ink pen and 20 students wrote with a black ink pen. Which of the following is the ratio of students writing with a black pen to those writing with a blue pen?

   F. 4 to 3
   G. 3 to 4
   H. 15 to 20
   J. 4 to 7

5. Which of the following lists the correct order of operations to simplify this expression?

   $$3 - 6(2 + 4) + 4 \div 2$$

   A. add, multiply, divide, subtract, add
   B. multiply, divide, add, subtract
   C. add, subtract, multiply, divide
   D. add, multiply, subtract, divide, add

6. Roger is packing super balls for shipping. Each ball has a diameter of 3 inches. The following shows the dimensions of the box. What is the volume of the box?

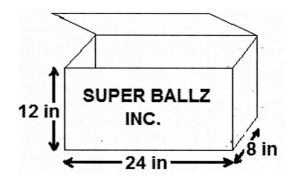

   F. $36 in^2$
   G. $44 in^2$
   H. $288 in^3$
   J. $2304 in^3$

The Green Thumb Club

The local plant and flower shop has a club for children between the ages of 5 and 9. Each child is given a tray of flower pots and seeds. Do numbers 7 through 10.

7. Noelle is saving money to buy a large flower pot to give to her mother. She has $14. After she sells her video game (v) she will have $21. Which of the following models this situation?

    A. $14 + 21 = v$
    B. $14 - v = 21$
    C. $v - 21 = 14$
    D. $14 + v = 21$

8. Each Saturday the children count the number of new sprouts in their flower pots. The number of new sprouts for eight weeks is listed below.

3, 6, 2, 9, 4, 6, 7, 2

What is the median number of new sprouts counted?

    F. 5
    G. 6
    H. 2
    J. 4.9

9. Kristen's first plant sprouted on the second Saturday after planting. She measured it each week when she went to the club meetings. She found that it grew 2.5 inches each week. Which of the following graphs best represents the height of the plant?

A

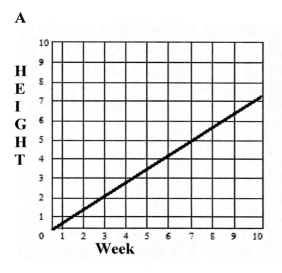

B

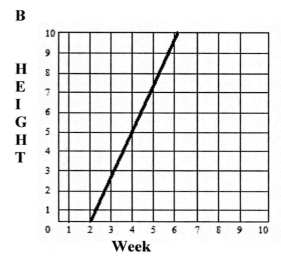

C

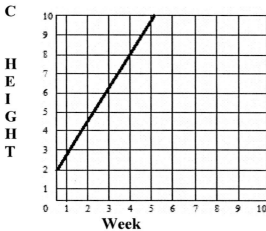

D

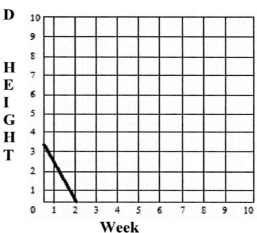

10. The children are going to build window boxes for their flowers. It takes four boards to build each box. They are going to build 20 boxes. The boards come in bundles of 3. How many bundles of boards will they need to order?

    F. 10
    G. 18
    H. 26
    J. 27

11. Solve: $3x + 7 = 19$

    A. 3
    B. -4
    C. 7
    D. 4

12. How many different outfits can be made from 3 pairs of pants, 3 shirts, and 4 belts?

    F. 4!
    G. 3 x 3 x 4
    H. 3 + 3 + 4
    J. 3!3!4!

13. Solve: $2(x - 3) - x + 2 = 1$

    A. -2
    B. 3
    C. -5
    D. 5

14. Solve: $4(x + 2) + 7 = 2x + 21$

    F. 3
    G. 7
    H. -7
    J. 0

15. Simplify: $\sqrt{64}$

    A. $\sqrt{8}$
    B. 8
    C. $4\sqrt{2}$
    D. 32

16. Evaluate: $x^3 + 2x^2 - 4x - 2$ **given** $x = -3$

    F. 1
    G. -27
    H. 10
    J. -12

17. Which of the graphs represents
$$y = \frac{2}{3}x + 2$$

A. 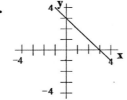   **B.**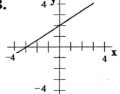

C. 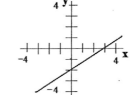   **D.**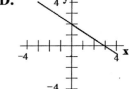

18. Vicki is buying sugar to make fudge. She can choose from the following brands:

Brand	Size	Price
Cain White	$\frac{1}{2}$ *pound*	$0.85
Great Value	3 *pounds*	$4.95
Pure Cain	5 *pounds*	$8.15

Which one of the following statements is true?

F. Pure Cain is the most expensive brand per pound.

G. Cain White and Great Value cost the same per pound.

H. Cain White is the most expensive brand per pound.

J. Great Value is the least expensive brand per pound.

19. How many feet of barbed wire will be needed to enclose the rectangular field shown in the drawing?

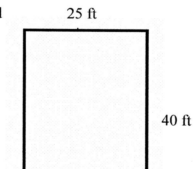

25 ft

40 ft

    A. 1,000 feet

    B. 65 feet

    C. 500 feet

    D. 130 feet

20. Denise earned the following scores on her tests in math class: 87, 91, 91, and 99. What is the mean of her test scores?

    F. 91

    G. 87

    H. 92

    J. 94

21. Jessica is installing decorative tile in her kitchen.

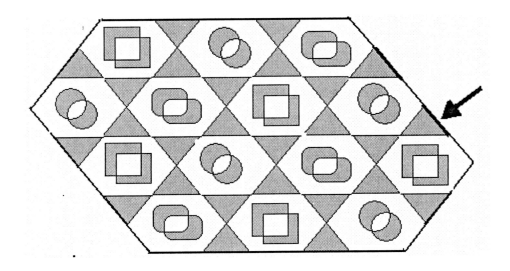

Which tile should be placed at the arrow to continue the pattern?

A.   **B.**   **C.**   **D.**

22. A graph contains the points (2, 1) and (-1, 3). Which of the following graphs

illustrates this line?

F.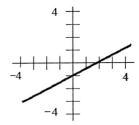

G.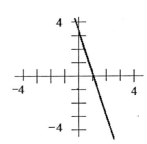

H.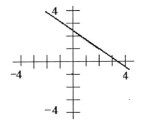

J.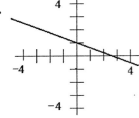

11

23. Which ordered pair is represented by the coordinates of point B shown on the graph?

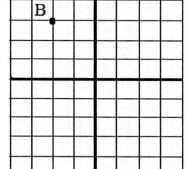

A. (2, 3)

B. (2, -3)

C. (-3, 2)

D. (-2, 3)

24. The pie chart shows the percentages of of students who prefer each of 5 soft drinks at a high school.

Half of the undecided students decided on Mongo Soda and the other half decided to choose Primo Soda. Which of the following statements is true?

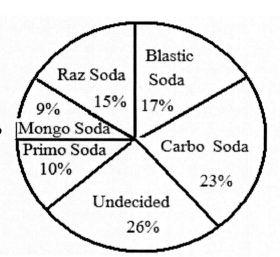

F. Carbo Soda will still be the most popular.

G. Blastic Soda will be the least popular.

H. Mongo Soda will be the most popular.

J. Primo Soda and Carbo Soda will be equally popular.

25. Which of the following linear graphs is represented by the table of values?

x	y
3	1
2	0
1	-1

A.

B.

C.

D.

26. Which shape would be next if the pattern were continued?

F.

G.

H.

J.

13

27. **Simplify:** $(6x^2 + 3x - 4) - (2x^2 + x - 3)$

    A. $(4x^2 + 2x - 1)$

    B. $(4x^2 + 4x - 7)$

    C. $(8x^2 + 4x - 7)$

    D. $(8x^2 - 4x + 1)$

28. **Solve:** $2(x - 3) = -2 - 5(x + 5)$

    F. 4

    G. -1

    H. $\dfrac{2}{3}$

    J. -3

29. Which one of the following functions generalizes the pattern in the table?

x	f(x)
-1	-1
0	1
3	7
5	11

    A. f(x) = x

    B. f(x) = 2x + 1

    C. f(x) = x − 1

    D. f(x) = 2x − 2

30. When J.B. attempts a putt, he succeeds 74.2% of the time. If J. B. attempts 24 putts, about how many times will he succeed?

    F. 98

    G. 18

    H. 9

    J. 6

31. **Evaluate:** $\dfrac{3}{4}x^2$ given x = 4

    A. 6

    B. 9

    C. 8

    D. 12

32. **Simplify:** $(2x + 1)(x - 3)$

    F. $3x - 2$

    G. $2x^2 - 3$

    H. $2x^2 - 5x - 3$

    J. $2x^2 + 7x - 3$

Bears

Directions

Every year park rangers count the number of bears living in the mountains. The table shows the results. Use the table to do numbers 33 through 38.

33. What was the median number of bear for

    the period shown in the table?

    A. 342

    B. 344

    C. 355

    D. 339

34. Which of the following expressions

    represents the average rate of change in

    the number of bears between 1990 and

    1995 in bears per year?

    F. $\dfrac{1995-1990}{347-341}$

    G. $\dfrac{347-341}{1995-1990}$

    H. $\dfrac{1990-341}{1995-347}$

    J. $\dfrac{341-347}{1995-1990}$

35. If the number of bears increases after 1995 at a rate of 7 bears per year, how many bears will there be in 2004?

    A. 417

    B. 396

    C. 354

    D. 410

36. In 1991 there were 158 female bears and 176 male bears. What was the ratio of male bears to female bears?

    F. 88 to 79

    G. 158 to 176

    H. 79 to 88

    J. 9 to 10

Year	Number of Bears
1990	341
1991	332
1992	325
1993	352
1994	355
1995	347

37. Which graph most accurately depicts the number of bears as a function of year?

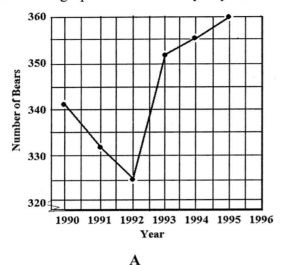

A

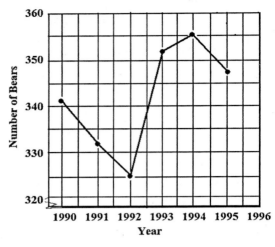

C

B

D

38. The number of bears a ranger saw on 5 days are listed below.

   5, 4, 10, 7, 4

What is the mean number of bears the ranger saw each day?

   F. 4

   G. 10

   H. 7

   J. 6

39. A school was having a bake sale. There were 400 chocolate chip cookies, 250 oatmeal cookies, and 300 sugar cookies. Which of the following is the correct ratio of chocolate chip cookies to oatmeal cookies?

    A. 3 to 4

    B. 4 to 3

    C. 4 to 7

    D. 1 to 4

40. Which of the following graphs represents $x > 2$?

    F.

    G.

    H.

    J.

41. Simplify: $(x+2)(x-2)$

    A. $x^2 - 4$

    B. $2x - 4$

    C. $x^2 + 4$

    D. $x^2 + 4x - 4$

42. Solve: $4(3x + 2) - 2x = 28$

    F. 2

    G. -2

    H. 4

    J. -4

43. Which of the following graphs shows a line with a slope of 0?

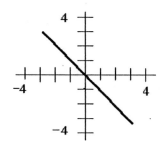

A

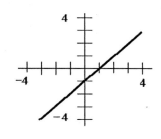

C

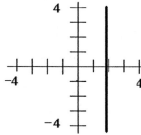

B

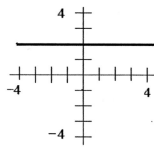

D

44. Which of these is the equation of the line that generalizes the pattern of the data in the table?

  F. $f(x) = x + 1$

  G. $f(x) = 2x + 1$

  H. $f(x) = 2x - 1$

  J. $f(x) = x - 1$

x	f(x)
0	1
1	3
2	5
3	7

45. A boy 5 feet tall casts a 2 foot shadow. At the same time a tree casts a 20 foot shadow. How tall is the tree?

  A. 50 feet

  B. 10 feet

  C. 40 feet

  D. 100 feet

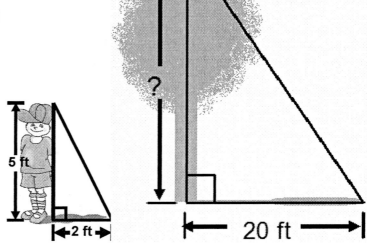

Note: Figures are not drawn to scale.

46. Which of these is the best estimate of the coordinate of Point P on the number line?

  F. $-2\dfrac{1}{4}$

  G. $-2\dfrac{3}{4}$

  H. $-1\dfrac{1}{4}$

  J. $-1\dfrac{3}{4}$

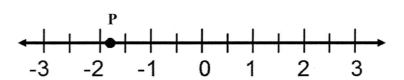

19

47. Each day the bakery at the local supermarket discounts the previous days pies by $1.00. The table shows the prices for selected pies.

Which of the following graphs represents the relation between the original price and sale price?

Pie	Original Price	Sale Price
Lemon	$4.00	$3.00
Cherry	$6.00	$5.00
Coconut	$10.00	$9.00
Fudge	$12.00	$11.00

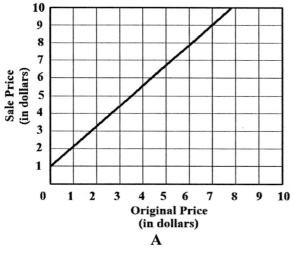

A

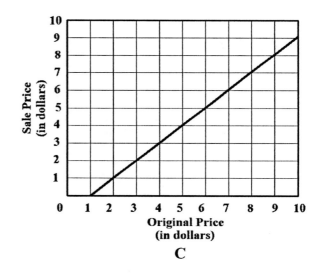

C

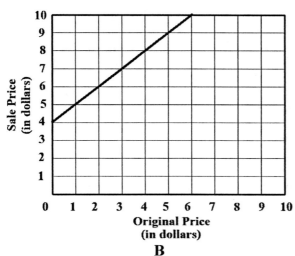

B

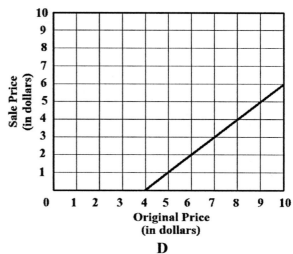

D

48. Estimate the area of the irregular figure shown on the grid.

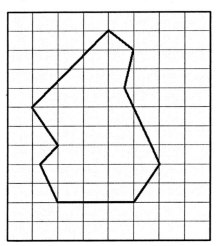

    F. 22

    G. 24

    H. 28

    J. 32

49. Which of the following graphs best represents the inequality $y \geq \frac{1}{2}x + 2$?

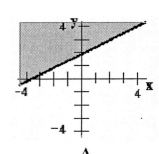

A

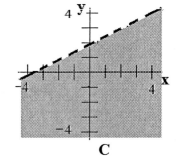

C

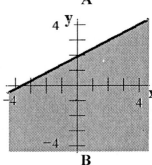

B

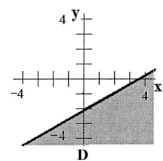

D

50. The relationship between inches and centimeters is shown on the graph below.

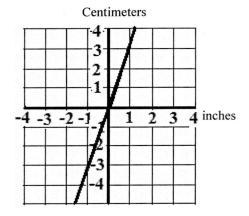

Which of the following is the best estimate of the number of centimeters in two inches?

    F. 7 cm

    G. 6 cm

    H. 5 cm

    J. 4 cm

51. Terence is buying a new tire for is truck. If the radius of his tire is 7 inches, what his it's circumference? (use $\frac{22}{7}$ for π)

    A. 22 inches

    B. 44 inches

    C. 96.34154 inches

    D. 256 inches

52. According to the diagram, what is the altitude of the balloon?

F. 30 ft

G. 50 ft

H. 80 ft

J. 120 ft

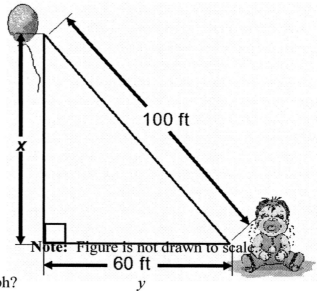

Note: Figure is not drawn to scale.

53. What is the slope of the line on the graph?

A. − 3

B. − 4

C. 3

D. 4

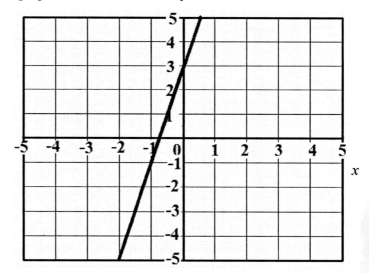

54. Which of these graphs represents $y = -3x + 2$?

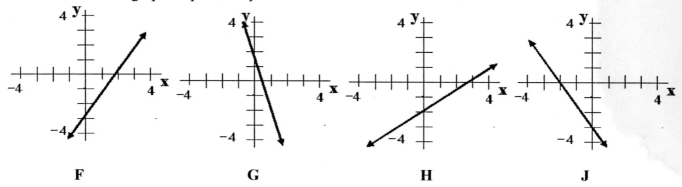

F   G   H   J

22

55. The size of a television is determined by measuring the screen diagonally. The picture shows the outer measurements of a television.

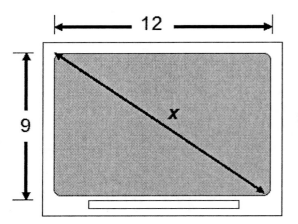

What is the size of the television?

    A. 10 inches

    B. 15 inches

    C. 27 inches

    D. 30 inches

56. The following graph represents the equation $y = x + 2$.

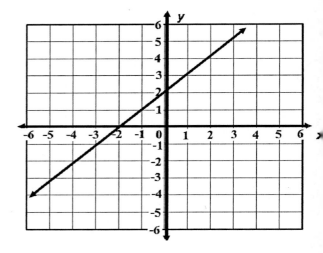

If the constant (2) changes from positive to negative, what will the graph look like?

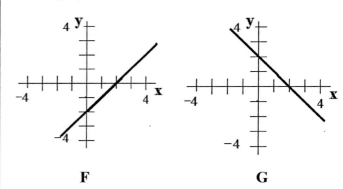

F   G

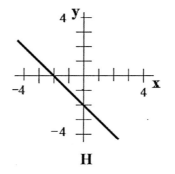

H   J

57. A gardener is mixing bags of potting mix. He makes two kinds. In the first kind he uses 8 pounds of compost and 6 pounds of fine sand. In the second kind he uses 3 pounds of compost and 4 pounds of fine sand. The gardener has 140 pounds of compost and 140 pounds of fine sand. He wants to make the maximum possible number of batches of potting mix and use as much of the compost and fine sand as he can. How many batches of each kind of mix can he make?

    A. 0 batches of the first and 30 batches of the second

    B. 10 batches of the first and 20 batches of the second

    C. 15 batches of the first and 15 batches of the second

    D. 20 batches of the first and 15 batches of the second

58. What is the reciprocal of 1?

    F. $\dfrac{1}{2}$

    G. -1

    H. $\dfrac{2}{1}$

    J. 1

59. Multiply: $(2x + 4)(3x - 6)$

    A. $6x - 24$

    B. $6x^2 - 24$

    C. $6x^2 - 2x + 24$

    D. $6x^2 - 2x - 24$

60. Workers are pouring a sidewalk around a rectangular garden. The outer perimeter of the sidewalk needs to be at least 78 feet. If the length of the sidewalk must be 14 feet longer than the width, what is the least possible integer value of the length of the sidewalk?

    F. 13 ft

    G. 18 ft

    H. 25 ft

    J. 27 ft

61. How many different ways can 5 different books be arranged on a shelf?

    A. 25

    B. 50

    C. 5!

    D. infinitely many

62. Which of the following figures shows an area representation of $(2x + 3)$ multiplied by $(x + 2)$?

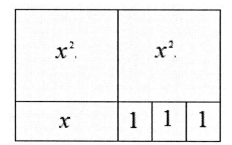

Practice Test 2

1. Estimate: 3.02(5.63)

    A. 8

    B. 15

    C. 17

    D. 21

2. Which of the following lists the correct order of operations to simplify the expression below?

$$3(4+6) - 2 \div 2$$

    F. multiply, divide, add, subtract

    G. add, multiply, divide, subtract

    H. multiply, add, subtract, divide

    J. multiply, add, divide, subtract

3. Tara is building a house. The house has a rectangular foundation. If the foundation measures 30 ft by 45 ft, how many square feet does the house have?

    A. 1350 sq ft

    B. 75 sq ft

    C. 150 sq ft

    D. 2700 sq ft

4. What number is represented by the point shown on the graph?

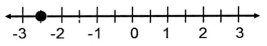

    F. 3

    G. 2.5

    H. -1.5

    J. -2.5

5. What is the reciprocal of 6?

    A. -6

    B. $-\dfrac{1}{6}$

    C. 6

    D. $\dfrac{1}{6}$

6. Jean can sew 3 buttons on a shirt in 5 minutes. How many buttons can she sew in 30 minutes?

    F. 18

    G. 36

    H. 45

    J. 90

7. Mr. Garcia has 18 students in his gardening club. Each of the students is going to build a miniature green house. It takes seven boards to build one green house. The boards come in packages of 5. How many packages will need to be ordered for all the students to be able to build a green house?

       A. 24

       B. 25

       C. 26

       D. 27

8. After the seeds have sprouted, they will be transplanted into a garden. The students are planning on building a fence around the garden. How much fencing will be needed to enclose a rectangular area measuring 10 ft by 25 ft?

       F. 35 ft

       G. 70 ft

       H. 250 ft

       J. 500 ft

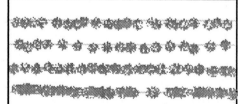

9. At the end of the second week the students counted the number of sprouts in each of their green houses. Below is a list of the number of sprouts for each of the green houses.

11, 13, 9, 12, 8, 7, 12, 10, 5, 8, 10, 11, 14, 7, 10, 9, 4, 6

What was the median number of plants counted?

A. 9
B. 9.5
C. 10
D. 9.2

10. Noelle is saving money to buy a larger green house. She has $18 in savings. After she receives her allowance (n), she will have $25. Which of the following equations models this situation?

F. $18 + 25 = n$
G. $25 - 18 = n$
H. $18 + n = 25$
J. $n - 25 = 18$

11. Kristen is growing a tomato plant from a seed. Each week she measures the plant to keep track of its growth. The plant sprouted on week 2. It grew 2 inches each week after that. Which graph best represents the growth of the tomato plants?

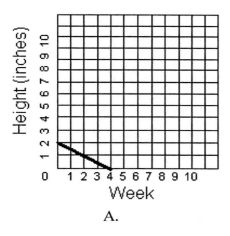

A.

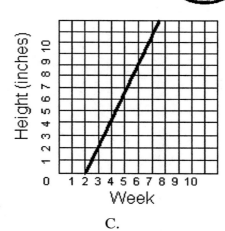

C.

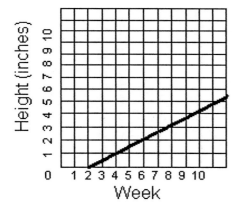

B.

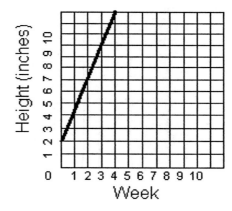

D.

12. Solve: $5x - 7 = 3$

    F. 2

    G. 3

    H. 5

    J. 7

13. How many different outfits can Kendra create from 5 shirts, 3 skirts, and 3 belts?

    A. $5 \times 3 \times 3$

    B. $5 + 3 + 3$

    C. $5!3!3!$

    D. $5!$

14. Solve: $2(2x + 4) - x = 5x - 4$

    F. ½

    G. 6

    H. 4

    J. ⅜

15. Solve: $4(x - 4) + 2x - 3 = 11$

    A. 3

    B. 4

    C. 5

    D. 6

16. Which of the following graphs represents $y = \frac{1}{3}x - 2$?

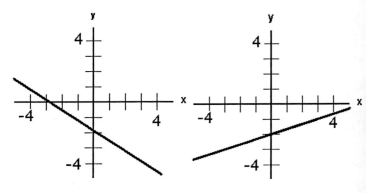

    F.   H.

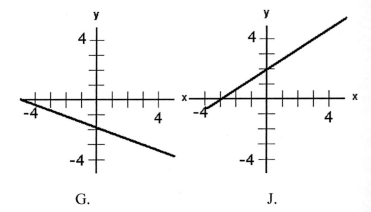

    G.   J.

17. Simplify: $\sqrt{256}$

    A. $\sqrt{16}$

    B. 256

    C. 16

    D. 8^2

18. Evaluate: $y = 2x^2 + 3x - 7$ for $x = -1$
    F. -8
    G. -2
    H. -1
    J. 11

19. Tydarius has test scores of 78, 84, 88, 74, and 91. What is the mean of his test scores?

    A. 80
    B. 82
    C. 83
    D. 88

20. Rakel is buying colored sand for an art project. She has the following brands to choose from.

Brand	Size	Price
SandArt	⅔ pound	$0.99
ColorSand	2 pounds	$2.35
Ocean Shades	5 pounds	$5.68

Which of the following statements is true?

    F. Ocean Shades is the most expensive brand
    G. SandArt costs more than ColorSand
    H. Ocean Shades costs more than SandArt
    J. ColorSand and Ocean Shades costs the same

21. Which is equal to $3(2+4) - \dfrac{4(2+3)}{5}$?

    A. 12

    B. 14

    C. 28

    D. -12

22. If one cut is made on a pizza, the maximum number of pieces that can be made is 2. If two cuts are made, the maximum is 4. If three cuts are made, the maximum is 7. What is the maximum number of pieces that can be made with four cuts?

    1   2   3

    2   4   7

    F. 10

    G. 11

    H. 14

    J. 21

23. Kwami repairs computers. He charges $15 plus $25 per hour for labor. Which equation gives the total charge for labor requiring h hours of work?

    A. $h + 15 = 2$

    B. $c = 15h + 25$

    C. $c = (15 + 25)h$

    D. $c = 15 + 25h$

24. Solve: $4c - 3 = -11$

    F. $-\dfrac{14}{4}$

    G. 2

    H. -2

    J. $\dfrac{4}{14}$

25. Beth's monthly cost for using an online tutoring service is given by the equation $d = 3(t - 3) + 6$, where d is the amount in dollars and t is the number of hours used. How many hours did she use the service if her bill was $15.00?

    A. 6 hours

    B. 8 hours

    C. 12 hours

    D. 25 hours

26. Which of the following represents the phrase "w squared increased by 5?"

    F. $2w + 5$

    G. $w^2 - 5$

    H. $w^2 + 5$

    J. $(w + 5)^2$

27. Estimate the area of the following shape.

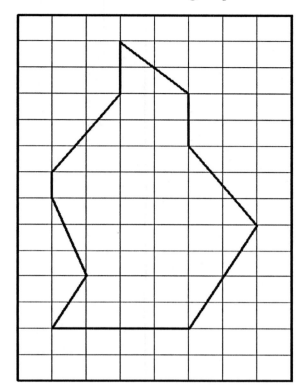

    A. 38 square units

    B. 40 square units

    C. 43 square units

    D. 48 square units

28. Tiara used 11 gallons of gasoline to drive 319 miles. Assuming she averages the same miles per gallon, how many gallons of gas will she need to drive 725 miles?

    F. 22

    G. 25

    H. 31

    J. 37

29. Taft High School is having a student government election. The pie chart shows the percentage of students expected to vote for each of the candidates. If half of the undecided students vote for Juan and the other half vote for Brad, which of the following statements is true?

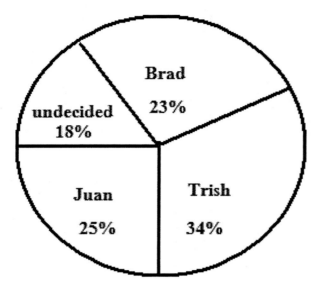

A. Juan will have the highest percentage of votes.

B. Trish will still be in the lead.

C. Juan and Trish will have the same percentage of votes.

D. Brad will have the highest percentage of votes.

30. About 60 million Americans have foot problems. Of those 60 million, about 6.1 million have flat feet or fallen arches. What percent of the people with foot problems have flat feet or fallen arches?

F. 6.1 %

G. 9.8%

H. 10.2%

J. 21.2%

31. What is 20% of 60?

    A. 10

    B. 12

    C. 15

    D. 30

32. Simplify:

$$(3x^2 - 4x + 2) - (x^2 - 6x + 6)$$

    F. $2x^2 - 10x + 8$

    G. $3x^4 - 10x^2 + 12$

    H. $2x^2 + 2x - 4$

    J. $4x^2 + 10x + 8$

33. One day while her class was on the playground, Ms. Williams noticed that some of her students had white tennis shoes and some had blue tennis shoes. There were 9 students with white shoes and 6 students with blue shoes. What is the ratio of students with blue shoes to those with white shoes?

    A. 9 to 6

    B. 3 to 2

    C 2 to 1

    D. 2 to 3

34. Simplify: $c \cdot d \cdot 2 \cdot c \cdot c \cdot 3 \cdot d \cdot c$

    F. $6c^4 d^2$

    G. $5 + 4c + 2d$

    H. $6 + c^4 + d^2$

    J. $6cd$

35. How much fencing would be needed to enclose a pasture 600 ft by 800 ft?

    A. 48000 ft

    B. 2800 ft

    C. 1400 ft

    D. 1000 ft

36. Solve and graph on the number line:

$$2(x + 1) < 8$$

F.

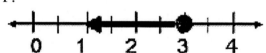

G.

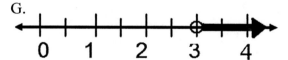

H.

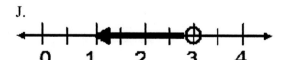

J.

37. Which graph is represented by the table of values?

x	y
-2	3
0	1
3	-2

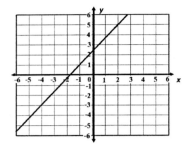

A

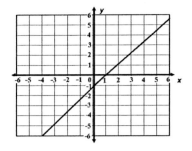

C.

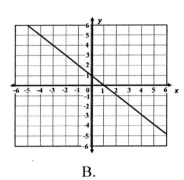

B.

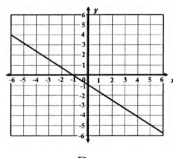

D.

38. Max is installing tile. To continue the pattern, what tile would be placed at the arrow?

F.

G.

H.

J.

37

ELEPHANTS

Each year, preservationists count the number of elephants in an African wildlife reserve. The table shows the results. Use the table to complete questions 39 – 44.

Year	Number of Elephants
1999	117
2000	121
2001	105
2002	99
2003	109
2004	111

39. What is the median number of elephants for the period shown in the table?

    A. 121

    B. 117

    C. 110

    D. 99

40. Which of the following expressions represent the average rate of change in the number of elephants between 1999 and 2004 in elephants per year?

    F. $\dfrac{1999-117}{2004-111}$

    G. $\dfrac{111-117}{1999-2004}$

    H. $\dfrac{2004-1999}{111-117}$

    J. $\dfrac{111-117}{2004-1999}$

41. In 2000 there were 55 male elephants and 66 female elephants in the reserve. What was the ratio of female to male elephants?

    A. 5 to 6

    B. 6 to 5

    C. 1 to 11

    D. 5 to 11

42. Which of these graphs most accurately depicts the number of elephants as a function of year?

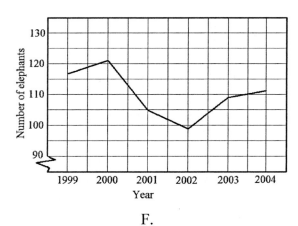

F.

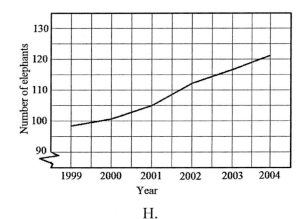

H.

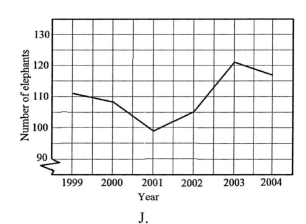

G.

J.

43. The number of elephants tourists saw on 5 days are listed below.

14, 18, 17, 14, 12

What is the mean number of elephants the tourists saw per day?

A. 14
B. 15
C. 17
D. 75

44. If the number of elephants increases after 2004 at a rate of 5 elephants per year, how many elephants will there be in 2010?

F. 116
G. 141
H. 146
J. 151

45. Nashville School of the Arts is having a dance in which 345 of the 415 students are planning on attending. What is the ratio of students planning to attend to the total population?

    A. 415 to 345

    B. 5 to 69

    C. 83 to 69

    D. 69 to 83

46. Simplify: $(x - 2)(x + 2)$

    F. $2x$

    G. x^2

    H. $x^2 + 4$

    J. $x^2 - 4$

47. Which is the best estimate of point P on the number line?

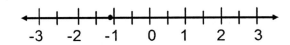

    A. $-1\frac{1}{8}$

    B. $-1\frac{1}{4}$

    C. $-1\frac{1}{2}$

    D. $-1\frac{3}{4}$

48. Which of the following is the equation of the line that generalizes the pattern of the data in the table?

x	f(x)
0	2
1	4
2	6
3	8

    F. $f(x) = x + 2$

    G. $f(x) = -x + 2$

    H. $f(x) = 2x + 2$

    J. $f(x) = 3x + 2$

49. Which point is shown on the graph?

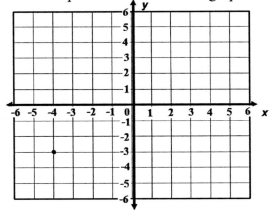

    A. (-4, -3)

    B. (4, 3)

    C. (-4, 3)

    D. (4, -3)

50. A statue 8 feet tall casts a 6 foot shadow. At the same time a flag pole casts a 9 foot shadow. How tall is the flag pole?

F. 11 feet

G. 12 feet

H. 16 feet

J. 18 feet

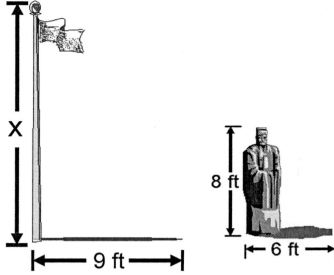

Note: Figures are not drawn to scale.

51. William wants to show his mother how well he can swim. He figured the farthest distance would be to swim from one corner diagonally to the far corner of the swimming pool. If the pool is 45 ft by 60 ft, how far did he swim?

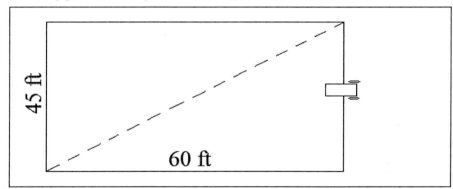

A. 60 ft

B. 75 ft

C. 105 ft

D. 5625 ft

52. Simplify: $(2x+2)(x-4)$

    F. $2x^2 - 6x - 8$
    G. $2x^2 + 6x - 8$
    H. $2x^2 - 8$
    J. $2x^2 - 4x + 8$

53. Evaluate: $\dfrac{3}{4}x^2$ given $x = 2$

    A. 1.50
    B. 3.00
    C. 4.25
    D. 14.50

54. When Ralph goes up to bat in baseball, he hits the ball 35.4% of the time. If he goes to bat 22 times, how many times will he hit the ball?

    F. 6
    G. 8
    H. 12
    J. 15

55. Solve: $2(x-3) = {}^-4 - 2(x-7)$

    A. -2
    B. -4
    C. 3
    D. 4

56. Which of the following graphs best represents the inequality $y \geq \dfrac{1}{2}x - 3$?

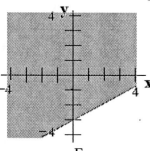

F.

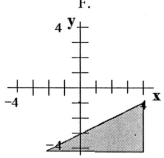

G.

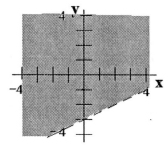

H.

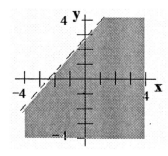

J.

42

57. According to the diagram, what is the altitude of the kite?

A. 40 ft

B. 60 ft

C. 80 ft

D. 180 ft

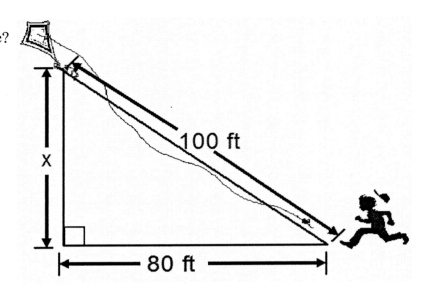

58. What is the slope of the line of the graph?

F. $\dfrac{1}{2}$

G. $\dfrac{2}{3}$

H. 2

J. 3

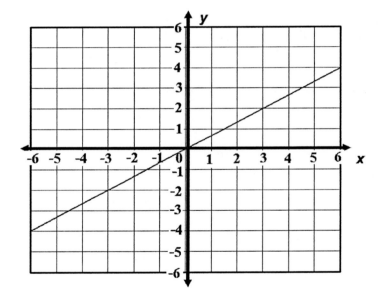

43

59. The outer parameter of a sidewalk around a monument needs to be at least 220 ft. If the length of the sidewalk must be 20 ft longer than the width, what is the least possible integer value of the length of the sidewalk?

A. 20 ft

B. 45 ft

C. 65 ft

D. 67 ft

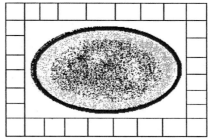

60. Multiply $(2x + 3)(3x - 6)$

    F. $5x - 3$ G. $6x^2 - 18$ H. $6x^2 + 3x - 18$ J. $6x^2 - 3x - 18$

61. Which of the following graphs represents the inequality $2x - y < 3$?

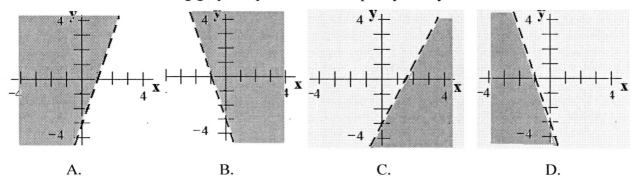

    A.   B.   C.   D.

62. Which of these figures is an area representation of $(2x + 3)$ multiplied by $(x + 2)$?

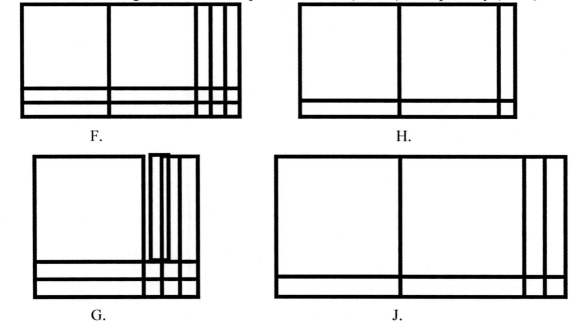

    F.   H.

    G.   J.

44

Practice Test 3

1. Simplify: $x \cdot x \cdot x \cdot y \cdot y \cdot y \cdot y$

    A. $3x + 4y$

    B. $x^3 y^4$

    C. $7xy$

    D. 12xy

2. Find the next number in the sequence.

    2, 4, 7, 11,

    F. 12
    G. 14
    H. 16
    J. 21

3. Which of the following sets of numbers are listed from least to greatest?

    A. 3.0, 3.4, 6, -7.3, -8.9, -9.1
    B. -4, 5, -7, 8, -10
    C. -4.1, -3, 1, 2.4, 4.2
    D. 3.2, 2.3, 0, -1.5, -4.3

4. On an English writing exam 18 students wrote with a blue ink pen and 27 students wrote with a black ink pen. Which of the following is the ratio of students writing with a black pen to those writing with a blue pen?

    F. 3 to 9
    G. 2 to 3
    H. 18 to 27
    J. 3 to 2

5. Which of the following lists the correct order of operations to simplify this expression?

$$2 - 4(8 + 4) + 6 \div 3$$

    A. add, multiply, divide, subtract, add
    B. multiply, divide, add, subtract
    C. add, subtract, multiply, divide
    D. add, multiply, subtract, divide, add

6. How much fencing would be required to build a fence around a rectangular dog pen measuring 25 feet by 35 feet?

    F. 875 ft
    G. 60 ft
    H. 120 ft
    J. 100 ft

The Green Thumb Club

The local plant and flower shop has a club for children between the ages of 5 and 9. Each child is given a tray of flower pots and seeds. Do numbers 7 through 10.

7. Noelle is saving money to buy a large flower pot to give to her mother. She has $16. After she sells her video game (v) she will have $24. Which of the following models this situation?

    A. $16 + 24 = v$
    B. $16 - v = 24$
    C. $v - 24 = 16$
    D. $16 + v = 24$

8. Each Saturday the children count the number of new sprouts in their flower pots. The number of new sprouts for eight weeks is listed below.

3, 4, 2, 8, 4, 6, 7, 3

What is the median number of new sprouts counted?

F. 2
G. 4
H. 6
J. 8

9. Kristen's first plant sprouted on the third Saturday after planting. She measured it each week when she went to the club meetings. She found that it grew 1.5 inches each week. Which of the following graphs best represents the height of the plant?

A
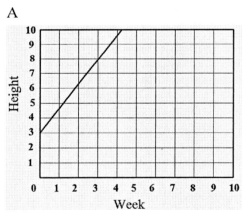

B
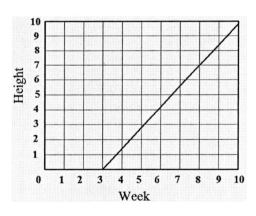

C
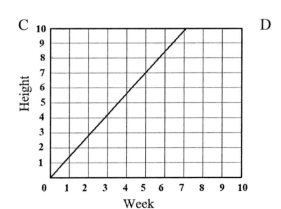

D
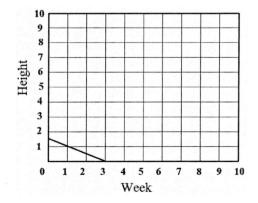

10. The children are going to build window boxes for their flowers. It takes four boards to build each box. They are going to build 25 boxes. The boards come in bundles of 7. How many bundles of boards will they need to order?

    F. 4
    G. 14
    H. 15
    J. 25

11. Solve: $3x + 7 = 19$

    A. 3
    B. -4
    C. 7
    D. 4

12. How many different outfits can be made from 5 pairs of pants, 3 shirts, and 2 belts?

    F. 5!
    G. 5 x 3 x 2
    H. 5 + 3 + 2
    J. 5!3!2!

13. Solve: $3(x - 3) - x + 4 = 1$

    A. -2
    B. 3
    C. -5
    D. 5

14. Solve: $3(x + 2) + 9 = 2x + 18$

    F. 3
    G. -3
    H. 4
    J. 0

15. Simplify: $\sqrt{81}$

    A. $\sqrt{9}$
    B. 9
    C. $3\sqrt{2}$
    D. 40.5

16. Evaluate: $x^3 + 2x^2 - 4x - 2$
    given: $x = -2$

    F. 0
    G. 1
    H. 6
    J. -12

17. Which of the graphs represents: $y = \frac{1}{2}x - 1$?

A.
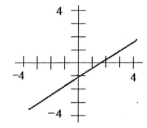

B.
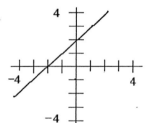

C.
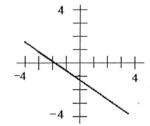

D.
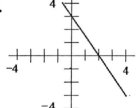

18. Vicki is buying sugar to make fudge. She can choose from the following brands:

Brand	Size	Price
Cain White	$\frac{1}{2}$ *pound*	$0.65
Great Value	3 *pounds*	$3.95
Pure Cain	5 *pounds*	$7.80

Which one of the following statements is true?

F. Pure Cain is the most expensive brand per pound.

G. Cain White and Great Value cost the same per pound.

H. Cain White is the most expensive brand per pound.

J. Great Value is the least expensive brand per pound.

19. How many feet of barbed wire will be needed to enclose the rectangular field shown in the drawing?

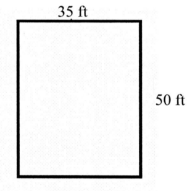

35 ft

50 ft

    A. 1,750 feet

    B. 85 feet

    C. 500 feet

    D. 170 feet

20. Denise earned the following scores on her tests in math class: 89, 94, 94, and 99. What is the mean of her test scores?

    F. 91

    G. 87

    H. 92

    J. 94

21. Jessica is installing decorative tile in her kitchen.

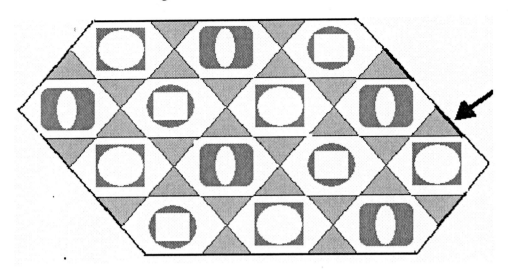

Which tile should be placed at the arrow to continue the pattern?

A.   **B.**   **C.**   **D.**

22. A graph contains the points (-3,-2) and (3, 1). Which of the following graphs illustrates this line?

F. 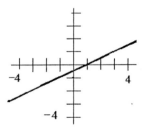   **G.**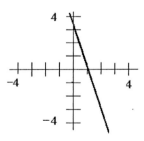

H. 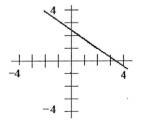   **J.**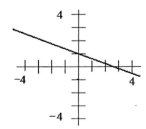

51

23. Which ordered pair is represented by the coordinates of point B shown on the graph?

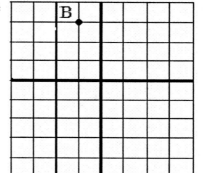

    A. (1, 3)

    B. (1, -3)

    C. (-3, 1)

    D. (-1, 3)

24. The pie chart shows the percentages of of students who prefer each of 5 soft drinks at a high school.

Half of the undecided students decided on Mongo Soda and the other half decided to choose Primo Soda. Which of the following statements is true?

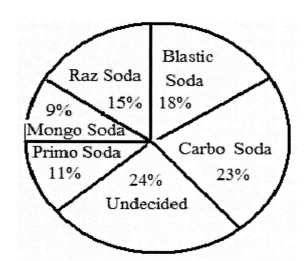

    F. Carbo Soda will still be the most popular.

    G. Blastic Soda will be the least popular.

    H. Mongo Soda will be the most popular.

    J. Primo Soda and Carbo Soda will be equally popular.

25. Which of the following linear graphs is represented by the table of values?

x	y
-2	0
0	2
1	3

A.

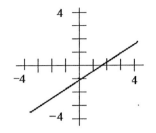

B.

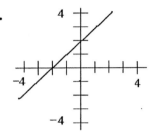

C.

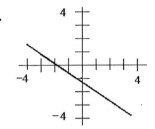

D.

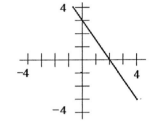

26. Maggie can buy different brands of sugar in different sizes as shown in the table.

Which one of the following statements is true?

Brand	Size	Price
Ultra White	4 pounds	$3.36
Pure Cain	8 pounds	$6.32
Cooker Elite	20 pounds	$17.60

    F. Ultra White is the least expensive per pound.

    G. Pure Cain is the most expensive per pound.

    H. Cooker Elite is more expensive per pound than Pure Cain.

    J. Cooker Elite is the least expensive per pound.

27. Simplify: $\left(3x^2 + 4x - 8\right) - \left(x^2 - 3x + 10\right)$

   A. $2x^2 + x + 2$

   B. $2x^2 + 7x - 18$

   C. $27x^3$

   D. $4x^2 + 7x + 18$

28. Solve: $3(x - 2) = -3 - 4(x + 5)$

   A. 3

   B. $\dfrac{7}{17}$

   C. $-\dfrac{17}{7}$

   D. -3

29. Which one of the following functions generalizes the pattern in the table?

x	f(x)
0	-1
1	0
2	1
3	2

   A. $f(x) = x$

   B. $f(x) = 2x + 1$

   C. $f(x) = x - 1$

   D. $f(x) = 2x - 2$

30. When J.B. attempts a putt, he succeeds 87.4% of the time. If J. B. attempts 24 putts, about how many times will he succeed?

    F. 90

    G. 21

    H. 9

    J. 3

31. Evaluate: $\frac{3}{4}x^2$ given x = 8

    A. 6

    B. 12

    C. 16

    D. 48

32. Simplify: $(3x + 4)(x - 2)$

    F. $3x^2 - 8$

    G. $4x + 2$

    H. $3x^2 - 2x - 8$

    J. $4x^2 - 2x + 8$

Bears

Directions

Every year park rangers count the number of bears living in the mountains. The table shows the results. Use the table to do numbers 33 through 38.

Year	Number of Bears
1990	269
1991	273
1992	281
1993	256
1994	269
1995	278

33. What was the median number of bears for the period shown in the table?

    A. 269

    B. 271

    C. 537

    D. 1626

34. Which of the following expressions represents the average rate of change in the number of bears between 1990 and 1995 in bears per year?

    F. $\dfrac{1995 - 1990}{278 - 269}$

    G. $\dfrac{278 - 269}{1995 - 1990}$

    H. $\dfrac{1990 - 269}{1995 - 278}$

    J. $\dfrac{269 - 278}{1995 - 1990}$

35. If the number of bears increases after 1995 at a rate of 7 bears per year, how many bears will there be in 2006?

    A. 417

    B. 396

    C. 355

    D. 310

36. In 1991 there were 142 female bears and 180 male bears. What was the ratio of male bears to female bears?

    F. 71 to 90

    G. 142 to 180

    H. 7 to 9

    J. 90 to 71

37. Which graph most accurately depicts the number of bears as a function of year?

A.
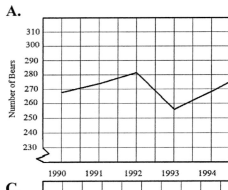

B.
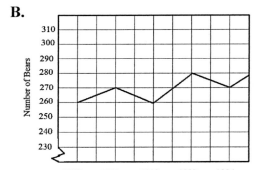

C.
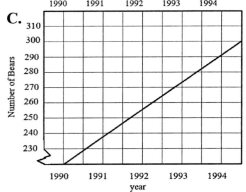

D.
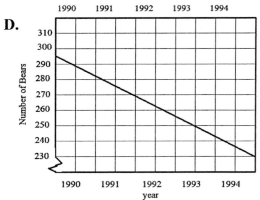

38. The number of bears a ranger saw on 5 days is listed below.

7, 4, 11, 8, 5

What is the mean number of bears the ranger saw each day?

    F. 4

    G. 10

    H. 7

    J. 6

39. A school was having a bake sale. There were 300 chocolate chip cookies out of a total of 1800 cookies. Which of the following is the correct ratio of chocolate chip cookies to the total number of cookies?

    A. 1 to 2

    B. 1 to 3

    C. 18 to 3

    D. 1 to 6

40. Which of the following graphs represents $x \leq 2$?

    F.
```
←+++++++++●+++++→
 -6 -5 -4 -3 -2 -1 0 1 2 3 4 5 6
```

    G.
```
←+++++++++●+++++→
 -6 -5 -4 -3 -2 -1 0 1 2 3 4 5 6
```

    H.
```
←+++++++++○+++++→
 -6 -5 -4 -3 -2 -1 0 1 2 3 4 5 6
```

    J.
```
←+++++++++○+++++→
 -6 -5 -4 -3 -2 -1 0 1 2 3 4 5 6
```

58

41. Simplify: $(x+3)(x-3)$

        A. $x^2 - 9$

        B. $2x - 9$

        C. $x^2 + 9$

        D. $x^2 + 4x - 4$

42. Solve: $3(2x + 2) - 2x = 22$

        F. 2

        G. -2

        H. 4

        J. -4

43. Which of the following graphs shows a line with an undefined slope ?

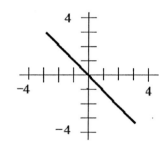

A

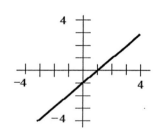

C

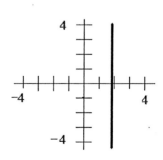

B

D

44. Which of these is the equation of the line that generalizes the pattern of the data in the table?

    F. $f(x) = x + 1$

    G. $f(x) = 2x + 1$

    H. $f(x) = 2x - 1$

    J. $f(x) = x - 1$

x	f(x)
0	-1
1	1
2	3
3	5

45. A boy 4 feet tall casts a 1 foot shadow. At the same time a tree casts a 20 foot shadow. How tall is the tree?

    A. 5 feet

    B. 16 feet

    C. 48 feet

    D. 56 feet

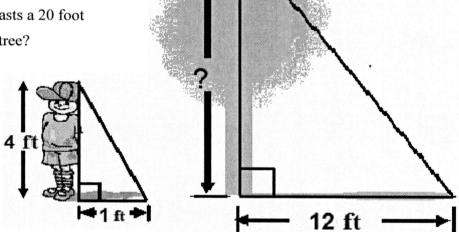

Note: Figures are not drawn to scale.

46. Which of these is the best estimate of the coordinate of Point P on the number line?

F. $-2\dfrac{1}{4}$

G. $-2\dfrac{3}{4}$

H. $-1\dfrac{1}{4}$

J. $-1\dfrac{3}{4}$

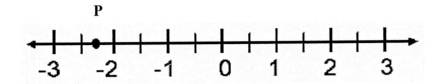

47. Each day the bakery at the local supermarket discounts the previous days pies by $2.00. The table shows the prices for selected pies.

Which of the following graphs represents the relation between the original price and the sale price?

Pie	Original Price	Sale Price
Lemon	$4.00	$2.00
Cherry	$6.00	$4.00
Coconut	$10.00	$8.00
Fudge	$11.00	$9.00

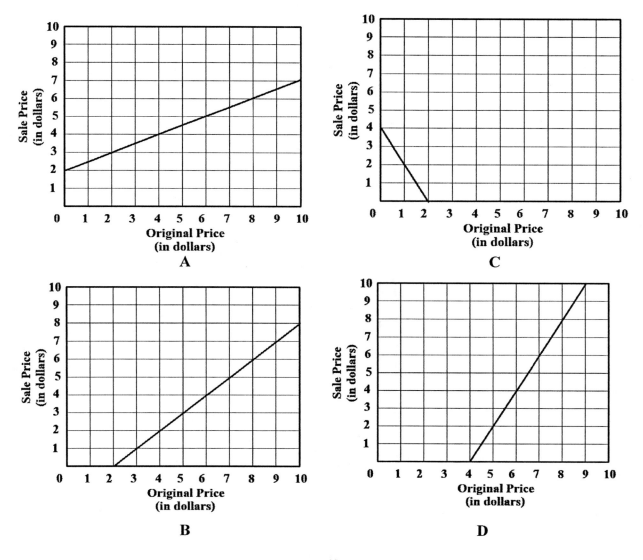

A

C

B

D

48. Estimate the area of the irregular
figure shown on the grid.

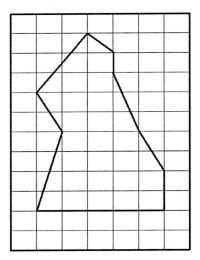

    F. 22

    G. 24

    H. 28

    J. 32

49. Which of the following graphs best

represents the inequality $y \le \dfrac{1}{2}x - 2$?

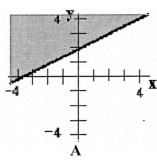

A

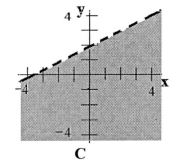

C

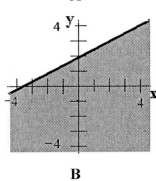

B

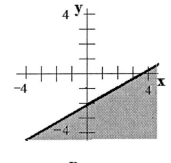

D

63

50. The relationship between inches and centimeters is shown on the graph below.

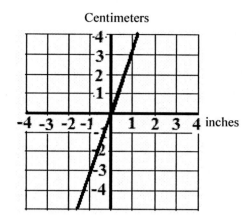

Which of the following is the best estimate of the number of centimeters in two inches?

    F. 7 cm

    G. 6 cm

    H. 5 cm

    J. 4 cm

51. Terence is buying a new tire for his truck. If the radius of his tire is 8 inches, what is its' circumference in terms of π?

    A. 8π inches

    B. 16π inches

    C. 64π inches

    D. 70π inches

52. According to the diagram, what is the altitude of the balloon?

F. 30 ft

G. 60 ft

H. 80 ft

J. 120 ft

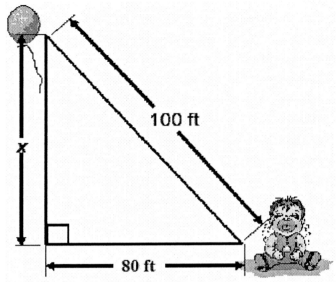

Note: Figure is not drawn to scale.

53. What is the slope of the line on the graph?

A. $\dfrac{2}{3}$

B. $-\dfrac{2}{3}$

C. $\dfrac{3}{2}$

D. $-\dfrac{3}{2}$

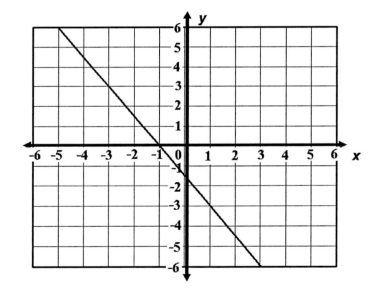

54. Which of these graphs represents $y = 2x - 3$?

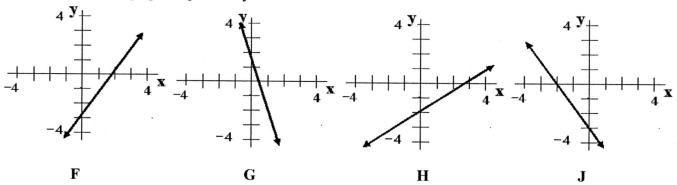

F   G   H   J

55. A contractor is planning to run an underground line diagonally across a field. The shortest distance is a straight line. How much line will he need to go across the field?

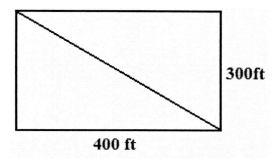

400 ft   **300ft**

A. 200 ft

B. 300 ft

C. 400 ft

D. 500 ft

56. The following graph represents the equation y = 2x + 2.

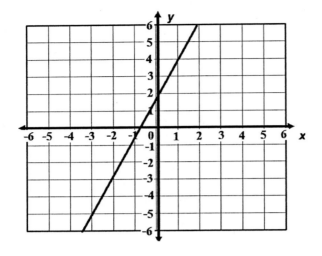

If the constant (2) changes from
positive to negative, what will the
graph look like?

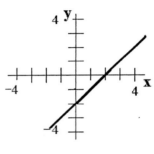

F

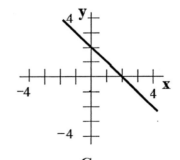

G

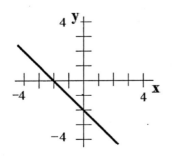

H

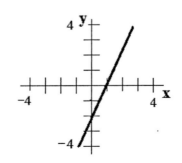

J

57. A gardener is mixing bags of potting mix. He makes two kinds. In the first kind he uses 4 pounds of compost and 3 pounds of fine sand. In the second kind he uses 6 pounds of compost and 8 pounds of fine sand. The gardener has 140 pounds of compost and 140 pounds of fine sand. He wants to make the maximum possible number of batches of potting mix and use as much of the compost and fine sand as he can. How many batches of each kind of mix can he make?

    A. 0 batches of the first and 30 batches of the second

    B. 20 batches of the first and 10 batches of the second

    C. 15 batches of the first and 15 batches of the second

    D. 15 batches of the first and 20 batches of the second

58. What is the reciprocal of 1?

    F. $\dfrac{1}{2}$

    G. -1

    H. $\dfrac{2}{1}$

    J. 1

59. Multiply: $(2x-3)(2x-5)$

    A. $4x+15$

    B. $4x^2-15$

    C. $4x^2-5x-15$

    D. $4x^2-16x+15$

60. Workers are pouring a sidewalk around a rectangular garden. The outer perimeter of the sidewalk needs to be at least 116 feet. If the length of the sidewalk must be 9 feet longer than the width, what is the least possible integer value of the length of the sidewalk?

    F. 34 ft

    G. 27 ft

    H. 25 ft

    J. 23 ft

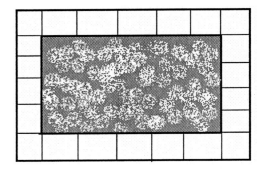

61. How many different ways can 7 books be arranged on a shelf?

    A. 49

    B. 70

    C. 7!

    D. infinitely many

62. Which of the following figures shows an area representation of (2x + 3) multiplied by (x +1)?

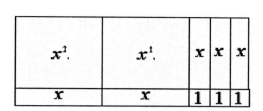

F

x^2	x^2	1
		1
		1
x	x	1
x	x	1
x	x	1

G

| x^2 | x^2 | | | | |
| 1 | 1 | 1 | 1 | 1 | 1 |

H

| x^2 | x^2 | | |
| x | 1 | 1 | 1 |

J

Explanations for Test 3

(problem) **15**

problem

⑮

1. Simplify: $x \cdot x \cdot x \cdot y \cdot y \cdot y \cdot y$

    A. $3x + 4y$

    B. $x^3 y^4$

    C. $7xy$

    D. $12xy$

In this problem, all of the x's and y's are being multiplied together. Since there are no addition signs in the problem, we can eliminate answer A as a possibility. Also, there are not any coefficients (numbers in front of letters) in the problem, so we can eliminate answers C and D. **That leaves B as the correct answer. (If there are numbers in the problem, just multiply the numbers together, and that answer goes in front.)**

Reason:

When multiplying variables, the exponents (the number at the upper right of the variable) are added together. If there is not an exponent showing, then the exponent is 1.

Example:

$x^2 \cdot x^5 = x^7$ That is the same as $(x^1 \cdot x^1) \cdot (x^1 \cdot x^1 \cdot x^1 \cdot x^1 \cdot x^1) = x^7$

In the first example, we add the 2 and the 5 to get 7. The second part of the example shows that x^2 is the same as $x \cdot x$. (I put in the 1's for the exponents so you could see that we are still just adding the exponents.)

2. Find the next number in the sequence.

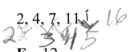

    2, 4, 7, 11, 16

    (#21)

    F. 12

    G. 14

    H. 16

    J. 21

To solve this one we are going to look at the spacing between the numbers.

To go from 2 to 4 we add 2, to go from 4 to 7 we add 3, to go from 7 to 11 we add 4.

Looking at these in order we added 2, then 3, then 4. To get the next number in the sequence we will add 5 to the 11 and get 16.

The correct answer is H.

3. Which of the following sets of
   numbers are listed from least to
   greatest?

   E. 3.0, 3.4, 6, -7.3, -8.9, -9.1
   F. -4, 5, -7, 8, -10
   G. -4.1, -3, 1, 2.4, 4.2
   H. 3.2, 2.3, 0, -1.5, -4.3

Whenever numbers are ordered from least to greatest, negative numbers go first. When dealing
with negative numbers, we look at them as being the opposite of positive numbers. In other
words, the bigger the negative number the smaller it is. -5 is smaller than -1.
In this problem, we can eliminate answers A and D because they have the positive numbers
listed first. We can also eliminate B because the numbers alternate between negative and
positive. **The correct answer is C.**

4. On an English writing exam 18 students
   wrote with a blue ink pen and 27 students
   wrote with a black ink pen. Which of the
   following is the ratio of students writing with a
   black pen to those writing with a blue pen?

   F. 3 to 9
   G. 2 to 3
   H. 18 to 27
   J. 3 to 2

It is important in this question to be sure and answer what is being asked. The question mentions
blue pens first then black pens. But we are supposed to give the ratio of black to blue. So, in our
ratio we have to list the black first then the blue.
The ratio is 27 black pens to 18 blue pens. To get our final answer we have to reduce the ratio, if
possible. Since 9 will go into 27 and 18, we divide both numbers by 9. Our final answer will be
3 to 2. **The correct answer for this question is J.**

5. Which of the following lists the correct
   order of operations to simplify this
   expression?

$$2 - 4(8 + 4) + 6 \div 3$$

   a. add, multiply, divide, subtract, add
   b. multiply, divide, add, subtract
   c. add, subtract, multiply, divide
   d. add, multiply, subtract, divide, add

Some students remember order of operations by the phrase **Please Excuse My Dear Aunt Sally** or **PEMDAS.** This stands for Parenthesis, Exponents, Multiplication, Division, Addition, and Subtraction.

First, we check to see if there are any parenthesis. Yes, so we work inside the parenthesis by adding the 8 and the 4 to get 12. Our new expression looks like this: $2 - 4(12) + 6 \div 3$. We can eliminate answer B since it says to multiply first.

Second, we check to see if there are any exponents. There are not any so we continue to the next step. We check to see if there is any multiplication and/or division. If there is we work from left to right. Yes, there is both multiplication and division so we perform those operations next. Our new expression looks like this: $2 - 48 + 2$. So, now we have added, multiplied and divided. Therefore, we can eliminate C and D, **which leaves A as the correct answer.** Our last step would be the addition and subtraction working from left to right. We would subtract 48 from 2 then add 2.

6. How much fencing would be required to build a fence around a rectangular dog pen measuring 25 feet by 35 feet?

    F. 875 ft
    G. 60 ft
    H. 120 ft
    J. 100 ft

To find the parameter of a rectangle we can use the formula on the reference page supplied for the Gateway exam. The formula for the parameter of a rectangle is $P = 2l + 2w$. l is the length and w is the width. Plugging in 25 for the width and 35 for the length we get: $P = 2(35) + 2(25)$. Using order of operations, we will multiply and then add. After multiplying, we get: $P = 70 + 50$. After adding, our final answer is 120 ft. **The correct answer is H.**

The Green Thumb Club

The local plant and flower shop has a club for children between the ages of 5 and 9. Each child is given a tray of flower pots and seeds. Do numbers 7 through 10.

7. Noelle is saving money to buy a large flower pot to give to her mother. She has $16. After she sells her video game (v), she will have $24. Which of the following models this situation?

      E. $16 + 24 = v$
      F. $16 - v = 24$
      G. $v - 24 = 16$
      H. $16 + v = 24$

Noelle is saving for the flower pot. She starts with $16 cash. When she sells her video game (v), the v in parenthesis means you are to use the letter v as the variable for the amount of money she gets for the video game. She will add that amount to the $16. We can write that as $16 + v$. We are told when she adds those two amounts together it will equal $24. **Therefore, our answer will be D**, $16 + v = 24$. The answer B does not make sense because that one subtracts the amount of the video game from her savings. Answer C does not make sense because it is subtracting the amount that she has after selling the video game from the amount of the video game. Answer A does not work because it is adding her savings to the amount of money she would have after selling the video game to get the price of the video.

8. Each Saturday the children count the number of new sprouts in their flower pots. The number of new sprouts for eight weeks is listed below.

3, 4, 2, 8, 4, 6, 7, 3

What is the median number of new sprouts counted?

    F. 2
    G. 4
    H. 6
    J. 8

We have to know that median means the middle number. Remember, the median is the middle of the road. Before finding the middle number, we must first arrange the numbers in order.
    2 3 3 4 4 6 7 8
If there is an even number of numbers, the middle will be between two numbers. When this happens, we take the average of the two numbers. In this case, the middle is between the two 4's. The average of 4 and 4 is 4. **The correct answer is G.**

9. Kristen's first plant sprouted on the third Saturday after planting. She measured it each week when she went to the club meetings. She found that it grew 1.5 inches each week. Which of the following graphs best represents the height of the plant?

A
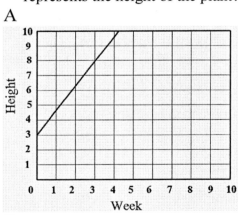

B
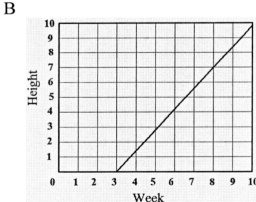

C
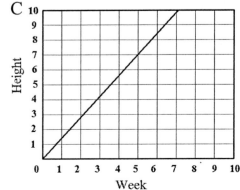

D
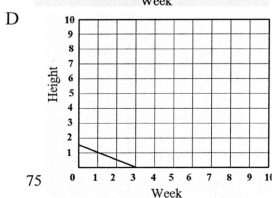

We are told that the plant sprouts on the third Saturday after being planted. Looking at the choices of the graphs, we need to find the graphs where the line starts on the third week. There are two, B and D. Graph A can be eliminated because that line starts out three inches high. Graph C can be eliminated because that shows the plant sprouting the day it was planted. When we look at graphs B and D, only B makes sense. Graph D actually shows the plant starting out at 1 ½ inches when it was planted and shrinking down to 0 inches on week three. **Graph B** shows the plant sprouting at week three and growing 1 ½ inches each week after that.

10. The children are going to build window boxes for their flowers. It takes four boards to build each box. They are going to build 25 boxes. The boards come in bundles of 7. How many bundles of boards will they need to order?

      I.   4
      J.   14
      K. 15
      J.   25

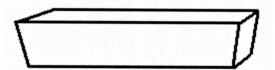

We need to know how many boards it will take to build 25 boxes. We are told that it takes 4 boards to build each box. Therefore, we need to multiply 25 times 4. It will take 100 boards to build the 25 boxes.

One way of solving this would be to divide 100 by 7. When we do the division, we get 14.2857…. In this case we have to round up to 15 because we have to have over 14 bundles. Another way of solving this one is to multiply 7 times each of the answers, starting with the smallest number, until we get to one that has at least 100 boards.

7 x 4 = 28 not enough

7 x 14 = 98 close, but still 2 boards short

7 x 15 = 105 we have enough to complete 25 boxes

It is ok to be a little over, but it is not ok to be a little short.
The correct answer is H.

11. Solve: $3x + 7 = 19$

     L. 3
     M. -4
     N. 7
     O. 4

One way of solving this problem is by using algebra.

$$3x + 7 = 19$$

     $\underline{-7 \quad -7}$ First, subtract 7 from both sides of the equal sign

     $\underline{3x} = \underline{12}$ After subtracting, our new equation is $3x = 12$

     3   3   To get the x by itself, divide both sides by 3

     $x = 4$ **The correct answer is D.**

Another way of solving this problem is substituting the answers in for x and using order of operations.

A. $3(3) + 7 = 19 \rightarrow 9 + 7 = 19 \rightarrow 16 = 19$ 16   **Does not equal 19, wrong answer**

B. $3(-4) + 7 = 19 \rightarrow -12 + 7 = 19 \rightarrow -5 = 19$ **Again , wrong answer**

C. $3(7) + 7 = 19 \rightarrow 21 + 7 = 19 \rightarrow 28 = 19$ **I hope the next one is correct**

D. $3(4) + 7 = 19 \rightarrow 12 + 7 = 19 \rightarrow 19 = 19$ **Correct**

12. How many different outfits can
be made from 4 pairs of pants,
3 shirts, and 2 belts?

     F. 5!
     G. 5 x 3 x 2
     H. 5 + 3 + 2
     J. 5!3!2!

Whenever we are confronted with a problem like this one, we just have to multiply all the numbers together. **The correct answer is G.** When you see a number with an explanation point next to it like 5!, that means to multiply 5 x 4 x 3 x 2 x 1. That is a factorial and is used in other types of problems.

13. Solve: $3(x - 3) - x + 4 = 1$

      P. -2
      Q. 3
      R. -5
      S. 5

Solving this one algebraically, we will start on the left side using the distributive property by multiplying 3 times x and 3 times -3.

$3(x) - 3(3) - x + 4 = 1$ → $3x - 9 - x + 4 = 1$

Next, we should combine our x's. $3x - x = 2x$. So our new equation is:

$2x - 9 + 4 = 1$

Next, we combine $- 9 + 4$ and get $- 5$. So, our new equation is:

$2x - 5 = 1$
$\underline{+5 \quad +5}$ Add 5 to both sides
$\dfrac{2x = 6}{2 \quad\quad 2}$ Divide both sides by 2

    $x = 3$ **The correct answer is B**

This is another problem that we could use the answers that are given, plug them in for x, and use order of operations.

A. $3((-2) - 3) - (-2) + 4 = 1$ Inside the parenthesis $-2 - 3 = -5$

   $3(-5) - (-2) + 4 = 1$ When we have $- (-2)$, we know two negatives make a positive

   $3(-5) + 2 + 4 = 1$ Next, we multiply $3(-5)$

   $-15 + 2 + 4 = 1$ Then, we add and subtract from left to right

       $-8 = 1$ Wrong answer

B $3((3) - 3) - (3) + 4 = 1$ Inside the parenthesis $3 - 3 = 0$

   $3(0) - (3) + 4 = 1$ When we have $- (3)$, just means $- 3$

   $3(0) - 3 + 4 = 1$ Next, we multiply $3(0)$

   $0 - 3 + 4 = 1$ Then, we add and subtract from left to right

       $1 = 1$ Correct answer

14. Solve: $3(x + 2) + 9 = 2x + 18$

    F. 3
    G. -3
    H. 4
    J. 0

Again, on this problem our first step will be the distributive property. We will multiply the 3 that is on the outside of the parenthesis by both the x and the 2.

    $3(x) + 3(2) + 9 = 2x + 17 \rightarrow 3x + 6 + 9 = 2x + 18$

Next, we need to combine the $+4 + 9$ on the left side.

$3x + 15 = 2x + 18$	
$-2x \qquad\quad -2x$	To get the x's on the same side, we subtract 2x from each side
$x + 15 = 18$	$3x - 2x$ leaves 1 x on the left, $2x - 2x$ leaves no x's on the left
$-15 \; -15$	To get the x by itself, we subtract 15 from each side of the equal sign
$x = 3$	**The correct answer is F**

Again, with this problem we can use the answers given to us to plug in for x and see if the two sides are equal.

A. $3((3) + 2) + 9 = 2(3) + 18$ Work inside parenthesis first, $3 + 2 = 5$

    $3(5) + 9 = 2(3) + 18$ Next, we do the multiplication on each side

    $15 + 9 = 6 + 18$ The final step is the addition on each side

    $24 = 24$ The correct answer is A

15. Simplify: $\sqrt{81}$

    A. $\sqrt{9}$
    B. 9
    C. $3\sqrt{2}$
    D. 40.5

Using a calculator or the reference page supplied with the Gateway exam we can look up the square root of 81.

$\sqrt{81} = 9$. Another way of looking at this, we need to find a number we can multiply times itself to get 81. $9 \times 9 = 81$.
The correct answer is B.

16. Evaluate: $x^3 + 2x^2 - 4x - 2$
   given: $x = -2$

   F. 0
   G. 1
   H. 6
   J. -12

For this problem, we will put -2 in place of the x. The first step is to put parenthesis everywhere there is an x.

$$(\blacksquare)^3 + 2(\blacksquare)^2 - 4(\blacksquare) - 2$$

Next, we fill in the -2 inside each set of parenthesis.

$$(-2)^3 + 2(-2)^2 - 4(-2) - 2$$

If you are using a scientific or graphing calculator, you can enter it just like it is written. If you are working it out by hand, use order of operations. We cannot simplify inside the parenthesis, so we will use the exponents. -2 raised to the power of 3 is -8. Remember, -2 raised to the power of 3 means -2 x -2 x -2 = -8. Then, -2 raised to the power of 2 is 4. This one is positive 4 because, when multiplying, two negatives make a positive. Therefore, -2 x -2 = 4. So, our new problem looks like this:

$$-8 + 2(4) - 4(-2) - 2$$

Next, we will do our multiplication and division. 2 x 4 = 8 and -4 x -2 = 8, again two negatives, when being multiplied, equal a positive. Our problem now looks like this:

$$-8 + 8 + 8 - 2$$

Then, we add and subtract working from left to right. $-8 + 8 = 0$, $0 + 8 = 8$, $8 - 2 = 6$.
The correct answer is H.

17. Which of the graphs represents: $y = \frac{1}{2}x - 1$?

A.
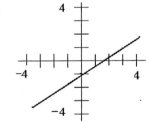

B.
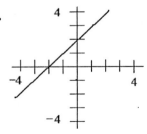

C.
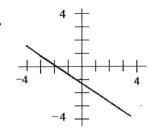

D.
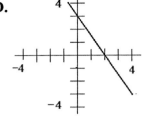

This equation is fairly easy to graph since it is in the form: $y = mx + b$. In this form, the number in front of the x, the number where the m is, is the slope. The b, which is our constant term, is the y-intercept. In other words, that is where the line crosses the y-axis. In this equation, the -1 is our y-intercept. So, we count down one place. If it were + 1, we would go up one place and put a point there.

The slope, the number in front of the x, is ½. Think of this as Rise over Run. Starting at the point on -1, we go up one place, then turn and go right 2 places and put our second point there. If the slope was negative, we would go down one place, then turn and go to the right 2 places. If the slope is positive, we go up. If it is negative, we go down, but we always go to the right.

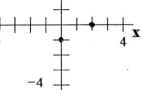

The last step is to draw a line through the two points to complete the graph.

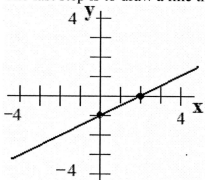

The correct answer is A.

18. Vicki is buying sugar to make fudge. She can choose from the following brands:

Brand	Size	Price
Cain White	$\frac{1}{2}$ *pound*	$0.65
Great Value	3 *pounds*	$3.95
Pure Cain	5 *pounds*	$7.80

Which one of the following statements is true?

F. Pure Cain is the most expensive brand per pound.

G. Cain White and Great Value cost the same per pound.

H. Cain White is the most expensive brand per pound.

J. Great Value is the least expensive brand per pound.

Our first step in solving this problem is to calculate how much each brand of sugar costs. We do this by dividing the price by the number of pounds.

$$\frac{price}{weight}$$

For Cain White we take $.65 \div \frac{1}{2}$ which is the same as .65 x 2. (When we divide a fraction, we flip the second one over and then multiply). This gives us $1.30 per pound.

For Great Value we take $3.95 \div 3 = 1.317$ which rounds off to $1.32 per pound. Last

is Pure Cain, we take $7.80 \div 5 = 1.56$ which is $1.56 per pound.

Cain White   = $1.30
Great Value = $1.32
Pure Cain   = $1.56

Now, we examine each of the statements to see which one is true.

F. Pure Cain is the most expensive brand per pound.

Yes, Pure Cain is the most expensive at $1.56 per pound. **The correct answer is F.**
None of the other statements are true.

19. How many feet of barbed wire will be needed to enclose the rectangular field shown in the drawing?

 35 ft

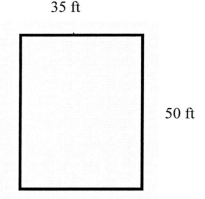

50 ft

    A. 1,750 feet

    B. 85 feet

    C. 500 feet

    D. 170 feet

To find the parameter of a rectangle we can use the formula on the reference page supplied for the Gateway exam. The formula for the parameter of a rectangle is $P = 2l + 2w$. l is the length and w is the width. Plugging in 35 for the width and 50 for the length we get: $P = 2(35) + 2(50)$. Using order of operations, we will multiply and then add. After multiplying, we get: $P = 70 + 100$. After adding, our final answer is 170 ft.

Another way of finding the parameter is adding all four sides together. Think of it as walking around the field. How far did you walk? Well, if I start in the top left corner and walk across the top, I have walked 35 feet. Then walking down the right side, I walk 50 feet. Walking across the bottom, I walk another 35 feet. Then finally, walking up the left side I walk another 50 feet. To determine how far I walked I add all the sides together.
$35 + 50 + 35 + 50 = 170$. **The correct answer is D.**

20. Denise earned the following scores on her tests in math class: 89, 94, 94, and 99. What is the mean of her test scores?

    F. 91

    G. 87

    H. 92

    J. 94

The mean is what most people think of as the average. Add up all the numbers, and divide by the number of tests. In this case we add:

$$89 + 94 + 94 + 99 = 376$$

Then we divide 376 by 4. $376 \div 4 = 94$ **The correct answer is J.**

21. Jessica is installing decorative tile in her kitchen.

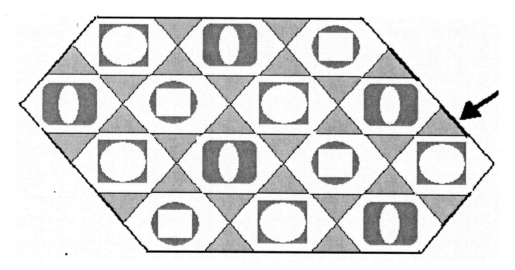

Which tile should be placed at the arrow to continue the pattern?

A.   **B.**   **C.**   **D.**

We can eliminate the triangle because there are no places where two sides of the triangles are together, because they all meet at the corners. Next, let's look at the pattern. The tiles repeat in the following order:

In the picture the arrow is pointing to the right of the tile that looks like this:  . In the

pattern, the tile to the right of that one is 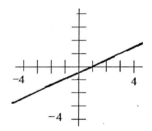 . **Therefore, the answer is C.**

22. A graph contains the points (-3,-2) and (3, 1). Which of the following graphs

illustrates this line?

F.

G.

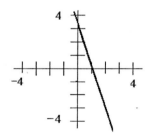

H.

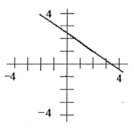

J.

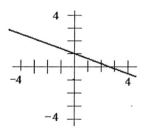

All we have to do on this one is graph the two points and draw a line through them. The first point is (-3, -2). Since the 3 is negative, we go back three places on the x axis. Then, since the 2 is negative, after going back three on the x axis we go down three from there and place a point in that spot.

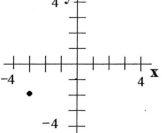

85

The second point is (3, 1). Since both of these are positive, we will go to the right and up. To graph this point we will go 3 places to the right on the x axis, then go up 1 place from there and place our point.

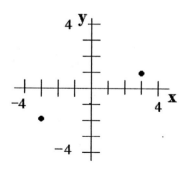

Now that we have our two points, we just draw a line through them.

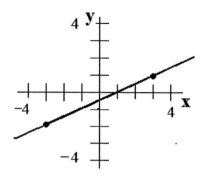

Our graph looks most like F.

23. Which ordered pair is represented by the coordinates of point B shown on the graph?

   A. (1, 3)

   B. (1, -3)

   C. (-3, 1)

   D. (-1, 3)

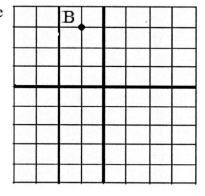

To get to this point from the center we have to go to the left one place and up three places. When we are looking for the coordinates of a point, we always start in the center. We move along the x axis (go left or right). If we move left, it is negative. If we move to the right, it is positive. Then we go up or down (this gives us our second number). If we go up, the number is positive. If we go down, the number is negative. Since we go left one place, the first number is -1. Then, we go up three places so our second number is 3. Therefore, our point is (-1, 3). **The correct answer is D.**

24. The pie chart shows the percentages of
students who prefer each of 5 soft drinks
at a high school.

Half of the undecided students decided on Mongo
Soda and the other half decided to choose Primo
Soda. Which of the following statements is true?

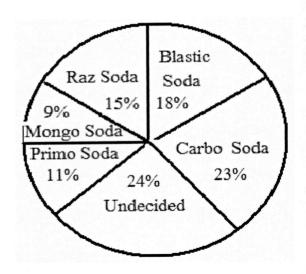

F. Carbo Soda will still be the most popular.

G. Blastic Soda will be the least popular.

H. Mongo Soda will be the most popular.

J. Primo Soda and Carbo Soda will be equally popular.

First, let's look at our new percentages after we give half of the undecided votes to Primo Soda and the other half to Mongo Soda. Our final percentages are as follows:

Raz Soda	15%
Blastic Soda	18%
Mongo Soda	21%
Primo Soda	23%
Carbo Soda	23%

Now, let's take a look at the statements.

F. Carbo Soda will still be the most popular.

At first, this may look correct. The problem is that now Primo Soda is tied with Carbo Soda. Therefore, Carbo Soda has to share the number one spot and is no longer the most popular.

G. Blastic Soda will be the least popular.

Blastic Soda has 18%, which is 3% more than Raz Soda. So Raz Soda is the least popular, not Blastic.

H. Mongo Soda will be the most popular.

Mongo Soda has 21% which is less than Primo Soda and Carbo Soda. So Mongo Soda is not the most popular.

J. Primo Soda and Carbo Soda will be equally popular.

Primo Soda and Carbo Soda each have 23%, so they are the same. **The correct answer is J.**

25. Which of the following linear graphs is represented by the table of values?

x	y
-2	0
0	2
1	3

A.

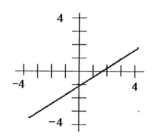

B.

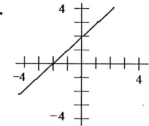

C.

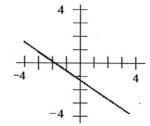

D.

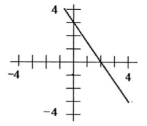

We simply plot the three points on a graph and draw a line through them. All three points should line up in a straight line. The first point is (-2, 0). That means we start in the center, we move 2 places to the left, and then we move 0 places up or down.

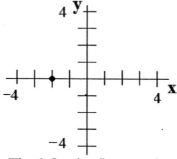

Our second point is (0, 2). The 0 for the first number means that we do not move left or right from the center. The 2 for the second number means we move up 2 places.

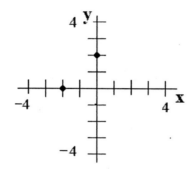

88

The third point is (1, 3). The 1 for the first number tells us to go to the right one place. The 3 in the second position means to go up 3 places and make our point.

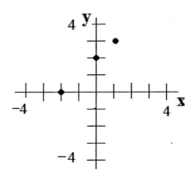

If we draw a line through the three points, it will look like answer B. **Therefore, the correct answer is B.**

26. Maggie can buy different brands of sugar in different sizes as shown in the table.

    Which one of the following statements is true?

Brand	Size	Price
Ultra White	4 pounds	$3.36
Pure Cain	8 pounds	$6.32
Cooker Elite	20 pounds	$17.60

    F. Ultra White is the least expensive per pound.

    G. Pure Cain is the most expensive per pound.

    H. Cooker Elite is more expensive per pound than Pure Cain.

    J. Cooker Elite is the least expensive per pound.

Our first step in solving this problem is to calculate how much each brand of sugar costs. We do this by dividing the price by the number of pounds.

$$\frac{price}{weight}$$

Ultra White $3.36 \div 4 = 0.84$

Pure Cain $6.32 \div 8 = 0.79$

Cooker Elite $17.60 \div 20 = 0.88$

Now we look at the statements given for the answers.

F. Ultra White is the least expensive per pound.

Ultra White is more expensive than Pure Cain, so this statement is not true.

G. Pure Cain is the most expensive per pound.

Both Ultra White and Cooker Elite are more expensive than Pure Cain, so this statement is not true.

H. Cooker Elite is more expensive per pound than Pure Cain.

Yes, Cooker Elite is more expensive per pound than Pure Cain. **The correct answer is H.**

27. Simplify: $\left(3x^2 + 4x - 8\right) - \left(x^2 - 3x + 10\right)$

    A. $2x^2 + x + 2$

    B. $2x^2 + 7x - 18$

    C. $27x^3$

    D. $4x^2 + 7x + 18$

To avoid making an arithmetic error, the easiest way to begin solving this problem is to change the subtraction sign to an addition sign, and change all the signs in the second set of parenthesis. In other words, we are going to change the minus sign in the middle and all the signs behind it. Our new problem looks like this:

$$\left(3x^2 + 4x - 8\right) + \left(-x^2 + 3x - 10\right)$$

Now we simply combine our like terms. First we have $3x^2 - x^2 = 2x^2$.

Our first term in our answer will be $2x^2$. Therefore we can eliminate answers C and D. For our second term we combine $4x + 3x = 7x$. Now we can eliminate answer A, which leaves us with B. We still want to make sure so we will combine our last two terms: $-8 - 10 = -18$. When using a calculator, type in the two numbers just the way you see them. Since the signs in front of both of the numbers are the same, we combine them. 8 plus 10 equals 18. Since both of the numbers have a minus, or negative, sign the answer is negative. So our final answer, after combining all our like terms is:

$$2x^2 + 7x - 18$$

The correct answer is B.

28. Solve: $3(x-2) = -3 - 4(x+5)$

    F. 3

    G. $\dfrac{7}{17}$

    H. $-\dfrac{17}{7}$

    J. -3

The first step in solving this equation is getting rid of the parenthesis on each side of the equal sign. On the left side we will multiply the 3 times the x and the 2. Since there is a minus sign in front of the 2, there will be a minus sign in front of the answer after multiplying by 3.

$$3(x - 2) = 3x - 6$$

Then, on the right side we will multiply the 4 times the x and the 5. Since there is a minus sign in front of the 4, we have to change the signs when we multiply. Remember, a negative times a positive is negative. (If the signs are different, the answer is negative. If the signs are the same, the answer is positive.)

$$- 4(x + 5) = -4x - 20$$

Our new equation is now:

$$3x - 6 = -3 - 4x - 20$$

Next, we combine the -3 and -20 on the right side of the equation.

$$-3 - 20 = -23$$

Now our equation has been simplified to this:

$$3x - 6 = -4x - 23$$

The next step is one in which many students have problems. We need to get all our x's on the same side of the equal sign. I prefer to move them to the side with the largest number of x's.

Since positive 3 is greater than negative 4, I will move the -4x to the left side. We will do this by adding 4x to both sides of the equal sign. (To move numbers or variables from one side to the other, we perform the opposite operation. The opposite of $-4x$ is $+4x$.)

$$
\begin{array}{r}
3x - 6 = -4x - 23 \\
+4x \qquad\quad +4x \\
\hline
7x - 6 = -23
\end{array}
$$

Now that we have all our x's on one side, we need to get all the numbers away from the x. We start with the number that is farthest away from the x on the same side of the equal sign. The -6 is farther away than the 7 so we will move the -6 first. The opposite of -6 is $+6$, therefore we will add 6 to both sides of the equal sign. (What we do to one side, we have to do to the other.)

$$
\begin{array}{r}
7x - 6 = -23 \\
+6 \quad\; +6 \\
\hline
7x = -17
\end{array}
$$

Our last step in solving this equation is dividing by 7. When we have a number next to a variable, it means that we are multiplying by that number. To move the 7, we need to do the opposite of multiplying which is dividing.

$$
\frac{7x}{7} = \frac{-17}{7}
$$

Our final answer is: $x = -\dfrac{17}{7}$. **Therefore, the correct answer is H.**

29. Which one of the following functions generalizes the pattern in the table?

x	f(x)
0	-1
1	0
2	1
3	2

A. $f(x) = x$

B. $f(x) = 2x + 1$

C. $f(x) = x - 1$

D. $f(x) = 2x - 2$

Don't let the f(x) confuse you in this problem. f(x) is simply another way of saying y. So, at the top of the table and in all the equations, where you see f(x), replace it with y.

x	y
0	-1
1	0
2	1
3	2

A. $y = x$
B. $y = 2x + 1$
C. $y = x - 1$
D. $y = 2x - 2$

The easiest way to solve one of these types of problems on a multiple choice test is to use the answers given in the problem. We will use the x and y they give us, substitute them into the equations, and see if the left side equals the right side.

A. $y = x$ Substitute 0 in for x and -1 in for y. We get: -1 = 0. -1 does not equal 0, so A is not the correct answer.

Now we will try substituting these numbers in for B.

B. $y = 2x + 1$ Substitute 0 in for x and -1 in for y. We get: $-1 = 2(0) + 1$. Solving this one, using order of operations, multiplication comes before addition, so we multiply 2 times 0 and get 0. -1 = 0 + 1, which is the same as $-1 = 1$. -1 does not equal 1, so B is not the correct answer. Now, we will try substituting these numbers in for C.

C. $y = x - 1$ Substitute 0 in for x and $- 1$ in for y. We get: $- 1 = 0 - 1$. 0 minus 1 equals $- 1$. So we get $- 1 = - 1$. -1 does equal -1. This could be the correct answer, but we need to check the others.

C. $y = x - 1$ Substitute 1 in for x and 0 in for y. We get: $0 = 1 - 1$. 1 minus 1 equals 0. So we get $0 = 0$. 0 does equal 0. So we keep going.

C. $y = x - 1$ Substitute 2 in for x and 1 in for y. We get: $1 = 2 - 1$. 2 minus 1 equals 1. So we get $1 = 1$. 1 does equal 1. We are pretty sure, at this point, that we have the correct answer, but we have to make sure.

C. $y = x - 1$ Substitute 3 in for x and 2 in for y. We get: $2 = 3 - 1$. 3 minus 1 equals 2. So we get $2 = 2$. 2 does equal 2. We have checked all of them, and they all work. **The correct answer is C.**

30. When J.B. attempts a putt, he succeeds

87.4% of the time. If J. B. attempts

24 putts, about how many times will

he succeed?

  F. 90

  G. 21

  H. 9

  J. 3

To solve this problem we simply multiply 87.4% times 24. Remember though, before we can multiply a percent times a number we have to change the percent to a decimal. To do this, divide 87.4 by 100 or simply move the decimal point two places to the left. Either way, we get .874. Now, we multiply .874 and 24.

$.874 \times 24 = 20.976$ which rounds to 21.

The correct answer is G.

31. Evaluate: $\frac{3}{4}x^2$ given $x = 8$

    A. 6

    B. 12

    C. 16

    D. 48

First, we substitute 8 in for the x in the problem. Now, we have:

$$\frac{3}{4}(8)^2$$

In the order of operations, we take care of our exponents before we divide.

$$(8)^2 = 64$$

Now, our problem looks like this:

$\frac{3}{4}(64)$ You can use a calculator to find the answer if you use parenthesis. Just enter the

problem like this: $(3 \div 4) \times (64)$ and press enter. If you are working it out step by step, you would multiply ¾ times 64. To do this we will put the 64 over 1 to make it into a fraction. Then, we can reduce and multiply straight across.

$\frac{3}{4} \times \frac{64}{1}$ Four will go into 4 and 64, so we can cross cancel to simplify the problem. (We have to use a top and bottom number to cross cancel.)

So now, our simplified problem looks like this:

$$\frac{3}{1} \times \frac{16}{1} = \frac{48}{1}$$

Since any number divided by 1 equals the number, our answer is 48.
The correct answer is D.

32. Simplify: $(3x + 4)(x - 2)$

    F. $3x^2 - 8$

    G. $4x + 2$

    H. $3x^2 - 2x - 8$

    J. $4x^2 - 2x + 8$

In this problem we are multiplying 2 binomials. In other words, we are multiplying two terms that each have two terms in them. When doing this type of problem, we multiply the first number in the first parenthesis by both of the numbers in the second parenthesis. Then, we multiply the second number in the parenthesis by both of the numbers in the second parenthesis, and then combine our two middle terms.

$$3x(x-2) = 3x^2 - 6x$$

$$4(x-2) = 4x - 8$$

Once we have completed both of the multiplications, we will line them up like this:

$$3x^2 - 6x + 4x - 8$$

Now, we combine our two middle terms. Remember to pay attention to the signs in front of the numbers. We have **− 6x + 4x** which equals **− 2x.** Now, our problem looks like this:

$$3x^2 - 2x - 8$$

The correct answer is H.

Directions

Every year park rangers count the number of bears living in the mountains. The table shows the results. Use the table to do numbers 33 through 38.

Year	Number of Bears
1990	269
1991	273
1992	281
1993	256
1994	269
1995	278

33. What was the median number of bears for the period shown in the table?

    A. 269

    B. 271

    C. 537

    D. 1626

Median is the middle number. Think about the median in the road. It is the middle of the road. Before we can determine the middle number, we must first line them up from smallest to largest.

$$256 \quad 269 \quad 269 \quad 273 \quad 278 \quad 281$$

The middle ends up being between the numbers 269 and 273. If the middle ends up between two numbers, take the average of the two numbers. We do that by adding the numbers together and dividing by 2.

$$\frac{269 + 273}{2} = \frac{542}{2} = 271$$

The correct answer is B.

34. Which of the following expressions represents the average rate of change in the number of bears between 1990 and 1995 in bears per year?

    F. $\dfrac{1995 - 1990}{278 - 269}$

    G. $\dfrac{278 - 269}{1995 - 1990}$

    H. $\dfrac{1990 - 269}{1995 - 278}$

    J. $\dfrac{269 - 278}{1995 - 1990}$

Rate of change is another way of saying slope. The slope formula is on your reference page that you may use on the Gateway.

Slope formula
$$\frac{y_2 - y_1}{x_2 - x_1}$$

When we are looking at a table, the left column is the x and the right column is the y. Therefore, the years are our x's and the number of bears are our y's. We can eliminate answer F because it has the number of bears on the bottom. We can eliminate answer H because there is a year and number of bears on both the top and bottom.

If we make our years and number of bears into points, with the year being x and the number of bears being y, we get **(1990, 269)** and **(1995, 278)**.

I will use (1990, 269) as my point 1, and (1995, 278) as my point 2.
$$x_1, y_1 \qquad\qquad x_2, y_2$$

Plugging the numbers into the formula we get:

$$\frac{278 - 269}{1995 - 1990}$$

The correct answer is G.

35. If the number of bears increases
    after 1995 at a rate of 7 bears per
    year, how many bears will there be
    in 2006?
    A. 417
    B. 396
    C. 355
    D. 310

There are two ways we can figure this one out. One way is to write down the years adding 7 for each one.

1996 1997 1998 1999 2000 2001 2002 2003 2004 2005 2006
  7 + 7 + 7 + 7 + 7 + 7 + 7 + 7 + 7 + 7 + 7 = 77
Now we add the 77 to the number of bears we had in 1995, which was 278.

$$77 + 278 = 355$$

The correct answer is C.

This way is ok because there is no time limit on the Gateway exams.

A quicker way of completing the problem is to subtract 1995 from 2006. Then multiply that answer times 7 since we increased by 7 bears per year.

$$2006 - 1995 = 11 \qquad 11 \times 7 = 77 \qquad 77 + 278 = 355$$

As you can see, we get the same answer either way.

The correct answer is still C.

36. In 1991 there were 142 female bears and 180 male bears. What was the

ratio of male bears to female bears?

    F. 71 to 90

    G. 142 to 180

    H. 7 to 9

    J. 90 to 71

We have to be careful on this one. The way it is written is kind of tricky. They tell us the number of females first then the number of males. Then, they want the ratio of males to females. So, on our ratio we have to list the number of males first then the number of females. It is the opposite of the way they are listed in the problem. We have 180 male bears and 142 female bears. Therefore, our ratio is:

180 to 142

Of course, this is not one of our choices because we are not finished. We have to simplify by dividing by our greatest common factor. The greatest common factor of 180 and 142 is 2. So, we need to divide 180 and 142 each by 2. Now we have:

90 to 71

The correct answer is J.

37. Which graph most accurately depicts the number of bears as a function of year?

A.

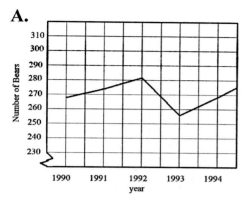

B.

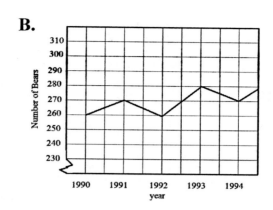

C.

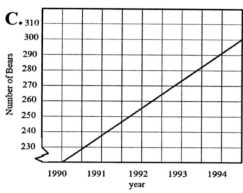

D.

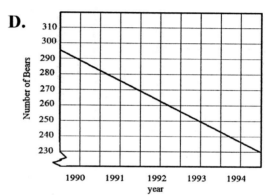

To solve this one we simply graph each of the
points. We will make a point above the year 1990 and across from 269. (Just a little below the 270 line.)
We will make a point above the year 1991 and across from 273. (Just a little above the 270 line.)
We will make a point above the year 1992 and across from 281. (Just a little above the 280 line.)
We will make a point above the year 1993 and across from 256. (Between the lines of 250 and 260.)
We will make a point above the year 1994 and across from 269. (Just a little below the 270 line.)
1995 is not on the graph.

When we plot the points our graph looks like this:

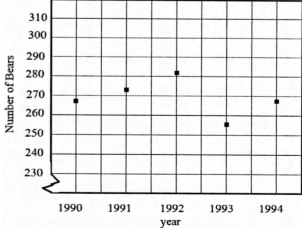

We connect the points from left to right, and our
graph looks like A.

The correct answer is A.

38. The number of bears a ranger saw on
5 days is listed below.

   7, 4, 11, 8, 5

What is the mean number of bears the
ranger saw each day?

   F. 4

   G. 10

   H. 7

   J. 6

The mean is what most people refer to as the <u>average</u>. We add the numbers together then divide
by however many numbers there are.

$$7 + 4 + 11 + 8 + 5 = 35 \qquad 35/5 = 7$$

There were 5 days, so divide 35 by 5 to get a mean of 7.

The correct answer is H.

39. A school was having a bake sale. There
were 300 chocolate chip cookies out of a
total of 1800 cookies. Which of the
following is the correct ratio of chocolate
chip cookies to the total number of
cookies?

   A. 1 to 2

   B. 1 to 3

   C. 18 to 3

   D. 1 to 6

We had 300 chocolate chip cookies out of a total of 1800 cookies in the bake sale. Our ratio is:

   300 to 1800

To get our final answer we have to divide each of the numbers by their greatest common factor.
The greatest common factor between 300 and 1800 is 300. When we divide each of the numbers
by 300 we get:

   1 to 6

The correct answer is D.

40. Which of the following graphs represents

$x \leq 2$?

F.

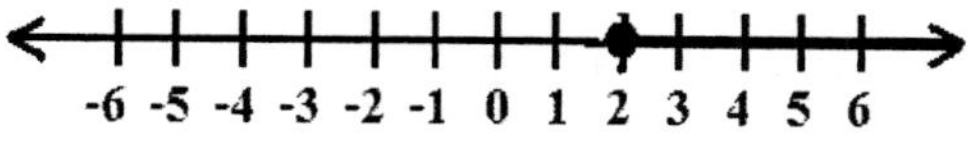

```
←++++++++●++++→
 -6 -5 -4 -3 -2 -1 0 1 2 3 4 5 6
```

G.

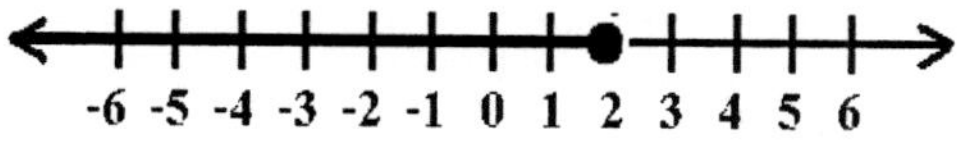

```
←++++++++●++++→
 -6 -5 -4 -3 -2 -1 0 1 2 3 4 5 6
```

H.

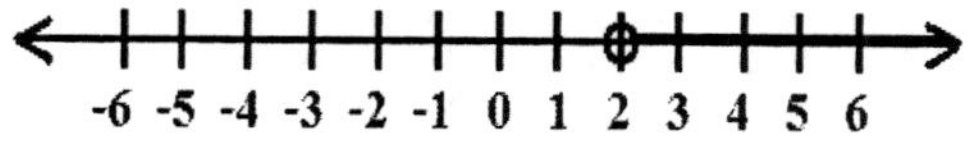

```
←++++++++○++++→
 -6 -5 -4 -3 -2 -1 0 1 2 3 4 5 6
```

J.

```
←++++++++○++++→
 -6 -5 -4 -3 -2 -1 0 1 2 3 4 5 6
```

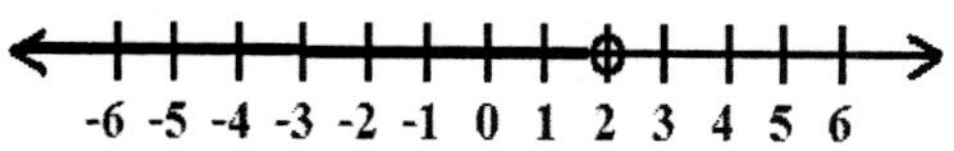

When graphing, if we have either of these symbols, $\leq$, $\geq$, we fill in the circle over the number. If we have either of these two symbols, $<$, $>$, we do not fill in the circle over the number. In other words, if there is a line under the symbol, fill it in. If there is not a line under the symbol, then don't fill it in.

In this problem we will fill in the circle over the point. Therefore, we can eliminate answers H and J since the circles are not filled in.

Now, we have to decide which direction to point the arrow. When we look at the symbol, if the little end is pointing at the x we go to the left. If the big end is pointing at the x, we go to the right. Remember, little left, big right. Since the little end is pointing at the x, we go to the left.

The correct answer is G.

41. Simplify: $(x + 3)(x - 3)$

   A. $x^2 - 9$

   B. $2x - 9$

   C. $x^2 + 9$

   D. $x^2 + 4x - 4$

In this problem we are multiplying 2 binomials. In other words, we are multiplying two terms that each have two terms in them. When doing this type of problem, we multiply the first number in the first parenthesis by both of the numbers in the second parenthesis. Then, we multiply the second number in the parenthesis by both of the numbers in the second parenthesis and then combine our two middle terms.

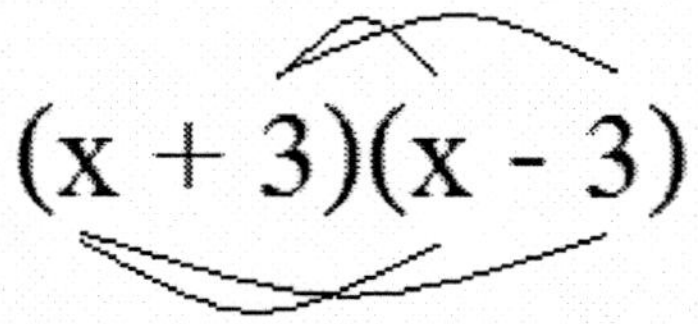

$$(x + 3)(x - 3)$$

102

We will multiply the two x's together and get x^2. Then, multiply the x in the first set of parenthesis times the -3 and get $-3x$. So, we now have $x^2 - 3x$ after multiplying both terms by x.

Next, we will multiply the 3 in the first set of parenthesis times the x and -3. When we multiply 3 times x, we get 3x. When we multiply 3 times -3, we get -9. After multiplying the 3 times both terms, we have: $3x - 9$.

Once we have completed both of the multiplications, we will line them up like this:

$$x^2 \boxed{-3x + 3x} - 9$$

Now, we combine the two like terms in the box. $-3x + 3x = 0$. The negative 3x and positive 3x cancel each other out. So, we are left with:

$$x^2 - 9$$

This type of problem is known as a difference in squares.

Many of you were probably taught to **Foil** the two terms. Foil stands for **first, outer, inner,** and **last.** Here's how it works.

First	**Outer**	**Inner**	**Last**
Multiply the first numbers together	Multiply the two outer numbers	Multiply the two inside numbers	Multiply the last two numbers
$(\mathbf{x} + 3)(\mathbf{x} - 3)$	$(\mathbf{x} + 3)(\mathbf{x} - \mathbf{3})$	$(\mathbf{x} + \mathbf{3})(\mathbf{x} - 3)$	$(\mathbf{x} + \mathbf{3})(\mathbf{x} - \mathbf{3})$
$(x)(x) = x^2$	$(x)(-3) = -3x$	$(3)(x) = 3x$	$(3)(-3) = -9$

Next, we list our answers from left to right. Be sure if there is a negative sign in front of the number that you bring it with the number in the problem.

$$x^2 - 3x + 3x - 9$$

Now, we combine the two like terms in the box. $-3x + 3x = 0$. The negative 3x and positive 3x cancel each other out. So, we are left with:

$$x^2 - 9$$

The correct answer is A.

42. Solve: $3(2x + 2) - 2x = 22$

        F. 2

        G. -2

        H. 4

        J. -4

The first step in solving this problem is to get rid of the parenthesis. We are going to do this by using the distributive property. In other words, we are going to multiply the 3 on the outside of the parenthesis by the 2x and the 2.

$$3(2x + 2) - 2x = 22$$

3 times 2x equals 6x, and 3 times 2 equals 6. So our new problem, without the parenthesis, looks like this:

$$6x + 6 - 2x = 22$$

Before we start moving the numbers away from the x, we need to simplify the left side some more. We can combine our 6x and – 2x. **6x – 2x = 4x.** So, now our equation looks like this:

$$4x + 6 = 22$$

Again, moving the numbers away from the x, we start with the number that is farthest away from the x. We need to move the 6 before we move the 4. (The 22 is farthest away from the x, but it is already on the other side of the equal sign, so we do not need to move it.) To move the 6 we perform the opposite operation. The opposite of + 6 is – 6, so we will subtract 6 from both sides.

$$\begin{array}{r} 4x + 6 = 22 \\ -\ 6 \quad -\ 6 \\ \hline 4x = 16 \end{array}$$

When we subtract the 6 from both sides, the + 6 and – 6 on the left cancel out, so we are left with the 4x on that side. Then, on the right side, 22 – 6 = 16.

Next, we need to move the 4. On the left side of the equation we have 4x. 4x actually means 4 times x. The opposite of multiplication is division; therefore, we will divide both sides by 4 to get the x by itself.

$$\frac{4x}{4} = \frac{16}{4}$$

When we divide 4x by 4 we are just left with the x on the left. 16 divided by 4 is 4. So, our final answer is:

$$x = 4$$

The correct answer is H.

Another way of solving this type of problem is to use the answers that are given and put them in place of x to see which one works.

F. 2 → $3(2(2) + 2) - 2(2) = 22$

Now, we perform the order of operations to solve the problem. First, we work inside the parenthesis.

$3(4 + 2) - 2(2) = 22$ → $3(2) - 2(2) = 22$

2 times 2 gives us 4. Then, we add the 2 inside the parenthesis to the 4 and get 6.
Next, we do our multiplication. We multiply 3 times 6 and get 18. Then, we multiply 2 times 2 to get 4. Now we have:

$18 - 4 = 22$

$18 - 4 = 14$ Since 14 does not equal 22, F is not the correct answer.

Now, let's try H.

F. 2 → $3(2(4) + 2) - 2(4) = 22$

Now, we perform the order of operations to solve the problem. First, we work inside the parenthesis.

$3(8 + 2) - 2(4) = 22$ → $3(10) - 2(4) = 22$

2 times 4 gives us 8. Then, we add the 2 inside the parenthesis to the 8 and get 10.
Next, we do our multiplication. We multiply 3 times 10 and get 30. Then, we multiply 2 times 4 to get 8. Now, we have:

$30 - 8 = 22$

$30 - 8$ does equal 22, so that is our correct answer.

43. Which of the following graphs shows a line with an undefined slope ?

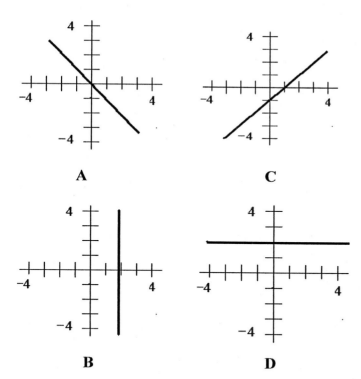

A

C

B

D

If a line is angled like answers A or B, it has a slope. Think of them like the side of a hill. The side of a hill has a slope, so we can narrow our choices down to B or D. It is probably easiest to identify a line with a slope of 0. The only line that we could put a 0 on without it rolling off is a 0 slope.

A zero slope line is horizontal. Therefore, the answer has to be B. An undefined slope is a vertical line. (It goes straight up and down.)

The correct answer is B.

44. Which of these is the equation of the line that generalizes the pattern of the data in the table?

    F. f(x) = x + 1

    G. f(x) = 2x + 1

    H. f(x) = 2x – 1

    J. f(x) = x – 1

x	f(x)
0	–1
1	1
2	3
3	5

Don't let the f(x) confuse you in this problem. f(x) is simply another way of saying y. So, at the top of the table and in all the equations, where you see f(x), replace it with y.

x	y
0	-1
1	1
2	3
3	5

F. $y = x + 1$

G. $y = 2x + 1$

H. $y = 2x - 1$

J. $y = x - 1$

This is just like problem #29. The easiest way to solve one of these type problems on a multiple choice test is to use the answers given in the problem. We will use the x and y they give us, substitute them into the equations, and see if the left side equals the right side.

F. $y = x + 1$ Substitute 0 in for x and – 1 in for y. We get: -1 = 0 + 1. -1 does not equal 0 + 1, so F is not the correct answer.

Now, we will try substituting these numbers in for G.

G. $y = 2x + 1$ Substitute 0 in for x and – 1 in for y. We get: - 1 = 2(0) + 1. Solving this one, using order of operations, multiplication comes before addition, so we multiply 2 times 0 and get 0. -1 = 0 + 1, which is the same as – 1 = 1. – 1 does not equal 1, so G is not the correct answer.

Now we will try substituting these numbers in for H.

H. $y = 2x - 1$ Substitute 0 in for x and – 1 in for y. We get: - 1 = 0 – 1. 0 minus 1 equals – 1. So, we get - 1 = - 1. -1 does equal -1. This could be the correct answer but we need to check the others.

H. $y = 2 x - 1$ Substitute 1 in for x and 1 in for y. We get: 1 =2(1) – 1. 2 x 1 = 2: 2 minus 1 equals 1. So, we get 1 = 1. 1 does equal 1. So we keep going.

H. $y = 2x - 1$ Substitute 2 in for x and 3 in for y. We get: 3 = 2(2) – 1. 2 x 2 = 4: 4 minus 1 equals 3. So, we get 3 = 3. 3 does equal 3. I am pretty sure, at this point, that we have the correct answer, but we have to make sure.

H. $y = 2x - 1$ Substitute 3 in for x and 5 in for y. We get: $5 = 2(3) - 1$. $2 \times 3 = 6$: 6 minus 1 equals 5. So, we get $5 = 5$. 5 does equal 5. We have checked all of them, and they all work.
The correct answer is H

45. A boy 4 feet tall casts a 1 foot shadow.
At the same time a tree casts a 12 foot
shadow. How tall is the tree?

    A. 5 feet

    B. 16 feet

    C. 48 feet

    D. 56 feet

80

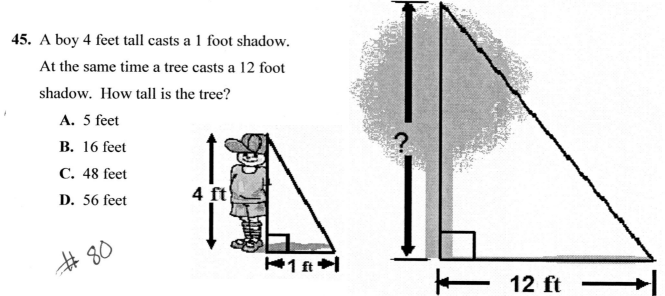

Note: Figures are not drawn to scale.

There are a couple of ways of solving this problem. One way is to simply compare the two triangles. The shadows are proportional to the height of the boy and the tree. We can ask ourselves what we have to multiply times 1 to get 12. The answer is, of course, 12. Now, we just multiply the boy's height times 12 to get the height of the tree.

$$4 \times 12 = 48$$

The correct answer is C.
Another way of solving the problem, the way math teachers want you to, is to set up a proportion. This will be 2 equal fractions. We will put the lengths of the shadow on top and the height of the boy and tree on the bottom of the fractions.

$$\frac{\text{Shadow}}{\text{Height}} \qquad \frac{1}{4} = \frac{12}{x}$$

We put an x in for the height of the tree because that is what we are looking for. Now, we are going to cross multiply and solve. We multiply 1 times x and that equals x. Next, we multiply 4 times 12 and that equals 48. So, our equation looks like this.

$$\frac{\text{Shadow}}{\text{Height}} \qquad \frac{1}{4} \diagup\!\!\!\!\diagdown \frac{12}{x} \qquad \rightarrow \quad x = 48$$

Again, the correct answer is C.

46. Which of these is the best estimate of the coordinate of Point P on the number line?

F. $-2\dfrac{1}{4}$

G. $-2\dfrac{3}{4}$

H. $-1\dfrac{1}{4}$

J. $-1\dfrac{3}{4}$

When we look at the graph, the point is between -2 and -3. We can eliminate answers C and D because those numbers are not between -2 and -3. The line between the -2 and -3 is -2.5 or -2 ½ . Since the point comes about half way between the -2 and the -2 ½ line, the point is closest to -2 ¼ .

The correct answer is F

47. Each day the bakery at the local supermarket discounts the previous day's pies by $2.00. The table shows the prices for selected pies.

Which of the following graphs represents the relation between the original price and the sale price?

Pie	Original Price	Sale Price
Lemon	$4.00	$2.00
Cherry	$6.00	$4.00
Coconut	$10.00	$8.00
Fudge	$11.00	$9.00

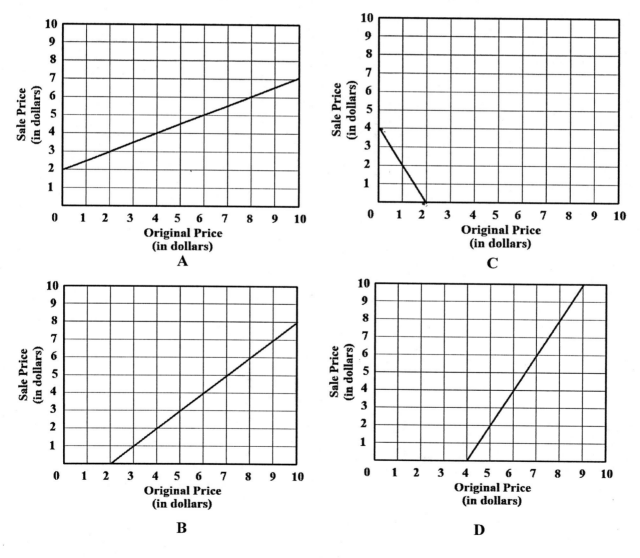

One way of solving this problem is to graph the points that are given. Remember, the first column is our x and the second column is our y.

For the first point, we will go the right 4 places (x axis) and up 2 places (y axis).
For the second point, we will go the right 6 places (x axis) and up 4 places (y axis).
For the third point, we will go the right 10 places (x axis) and up 8 places (y axis).

Two points are actually enough to give us our line. I personally prefer to use at least three to make sure they are in a straight line. When we graph our points, they line up like this:

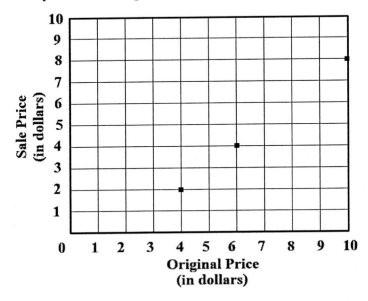

If we draw a line through the three points, it most closely matches answer B.

The correct answer is B.

48. Estimate the area of the irregular
    figure shown on the grid.

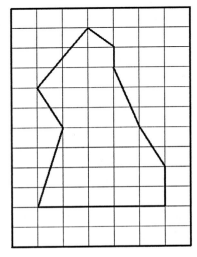

F. 22

G. 24

H. 28

J. 32

A lot of students have trouble with this type of problem. We just need to have a system of keeping track of the squares. First, let's count the total number of whole squares.

111

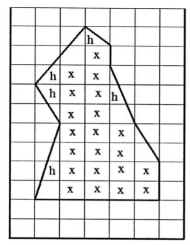

We have 21 complete squares. We don't have to worry about the squares with the x's anymore. Now, let's count the half squares or the ones that look pretty close to half by our best estimate.

We have 5 halves which makes 2 ½ wholes. We add the 2 ½ to 22 and get 24 ½ squares so far. We still have some counting to do. Now, we are going to count the squares that have more than half inside the area. We will not count the little pieces that are left because we are counting the squares that are not completely in the area.

We have 4 sections that have over half of the square inside the area. We add the 4 to the 24 ½ that we already have, and we get 28 ½ . **This is closest to answer H.**

We can solve this more accurately using the formulas from our reference page. The reference page gives us the formulas for figuring the area of rectangles and triangles.

Rectangle → $l \times w$ l is the length and w is the width

Triangle → $\dfrac{1}{2}bh$ b is the base (the longest side) h is the height

When we divide our figure into rectangles and triangles, it looks like this. (There is more than one way this can be done.)

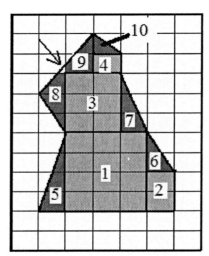

Numbers 1, 2, 3, and 4 are rectangles. A square is a rectangle, so we will use the formula for rectangles for those.

Numbers 5, 6, 7, 8, 9 and 10 are all triangles. The arrow points to a section where two triangles overlap. It is so small that we are not going to worry about it.

Number 1 is a 3 x 4 rectangle. 3 x 4 = 12 That rectangle has 12 squares in it.
Number 2 is a 1 x 2 rectangle. 1 x 2 = 2 That rectangle has 2 squares in it.
Number 3 is a 2 x 3 rectangle. 2 x 3 = 6 That rectangle has 6 squares in it.
Number 4 is a 1 x 1 rectangle. 1 x 1 = 1 That rectangle has 1 square in it.

Now, the triangles. On these, we will use the formula $\dfrac{1}{2}bh$.

Triangle 5 has a base of 5 and a height of 1. Using the formula we get $\dfrac{1}{2}(5)(1) = 2.5$.

Triangle 6 has a base of 2 and a height of 1. Using the formula we get $\dfrac{1}{2}(2)(1) = 1$

Triangle 7 has a base of 3 and a height of 1. Using the formula we get $\dfrac{1}{2}(3)(1) = 1.5$

Triangle 8 has a base of 3 and a height of 1. Using the formula we get $\frac{1}{2}(3)(1) = 1.5$

Triangle 9 has a base of 2 and a height of 1. Using the formula we get $\frac{1}{2}(2)(1) = 1$

Triangle 10 has a base of 1 and a height of 1. Using the formula we get $\frac{1}{2}(1)(1) = 0.5$

Now, we add all of our answers, from the rectangles and triangles, together to get our answer.

$$12 + 2 + 6 + 1 + 2.5 + 1 + 1.5 + 1.5 + 1 + 0.5 = 28$$

The correct answer is H.

49. Which of the following graphs best represents the inequality $y \leq \frac{1}{2}x - 2$?

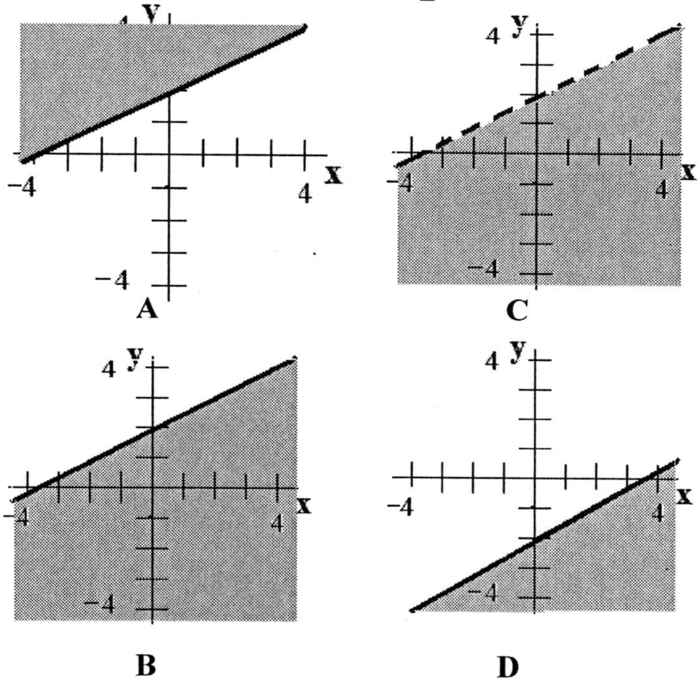

A

C

B

D

Before we worry about the shaded area, let's just graph the line. This one is easy because it is in slope intercept form which is y = mx + b. The m tells us our slope and the b tells us where it crosses the y axis. This line crosses the y axis at − 2. On my graph, I will count down 2 places from the center and make a point.

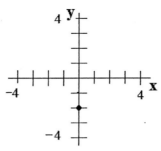

Now, using the number in front of the x, which is our slope, we can figure out where our second point belongs. When looking at the slope, think rise over run. The top number tells us how far up or down to go (if the slope is positive, go up, if negative, go down), and the bottom number tells us how far to go to the right (we always go to the right).

Starting at the point we graphed, we go up 1 and to the right 2.

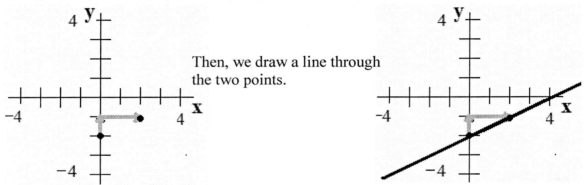

Then, we draw a line through the two points.

Now that we know where the line is, we can work with the inequality sign. With an inequality sign we have to determine if the line is going to be solid or dotted.
If the sign does not have the line under it, it is a dotted line.
Example: if we have either one of these two signs, < or >, it will be a dotted line.
If the sign has a line under it, we will draw a solid line.
Example: if we have either one of these two signs, ≤ or ≥, it will be a solid line.

Since the sign has a line under it we know the answer will have a solid line. So, we can eliminate answer C.

Answers A and B both have lines that cross the y-axis at positive 2. Our graph crosses the y-axis at negative 2. Also, the problem says that y is less than or equal (≤) to
½ x – 2. If the small end of the inequality sign is pointing at the y, shade under the line. If the open side of the inequality sign is pointing at the y, shade above the line.

The correct answer is D.

50. The relationship between inches and centimeters is shown on the graph below.

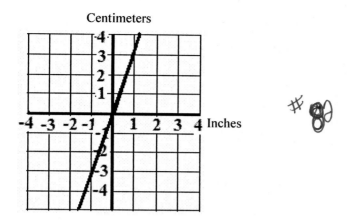

Which of the following is the best estimate of the number of centimeters in two inches?

    F. 7 cm

    G. 6 cm

    H. 5 cm

    J. 4 cm

To solve this problem, or any problem of this type with a graph, simply find the 2 on the line marked inches. Follow the line straight up until you get to the sloped line. Then, go left to see how many centimeters are on the centimeter axis. This one is a little different than most because the intersection is off the graph. In this case, you will have to extend the lines of the graph. Your graph should look like this:

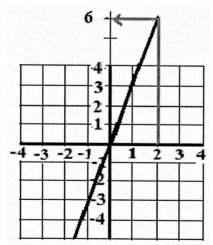

Going up from the 2, until we get to the sloped line, then going horizontally to the left, you should get 6 centimeters.

Another way of solving this problem is to find how many centimeters there are in one inch, and then double that amount.

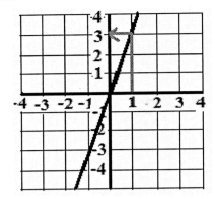

You find that 1 inch comes out to be about 3 centimeters. Therefore, 2 inches would be 6 centimeters. **The correct answer is G.**

51. Terence is buying a new tire for his truck. If the radius of his tire is 8 inches, what is it's circumference in terms of π?

    A. 8π inches

    B. 16π inches

    C. 64π inches

    D. 70π inches

On the reference page that you may use on the Gateway exam, there is a formula for calculating the circumference of a circle given the radius.

Circumference = $2\pi r$ In the formula r represents the radius. So, substituting 8 in for r, we get: $2\pi 8$. In our answers we see that there is only one number and the pi symbol. There are no addition or subtraction signs, so that means it is all multiplication. We multiply the 2 times 8 and get 16. (We can do that because of the commutative property of multiplication.) Don't worry though, about why, just know that it works. So our answer, after multiplying, is 16π.

The correct answer is B.

52. According to the diagram, what is the altitude of the balloon?

F. 30 ft

G. 60 ft

H. 80 ft

J. 120 ft

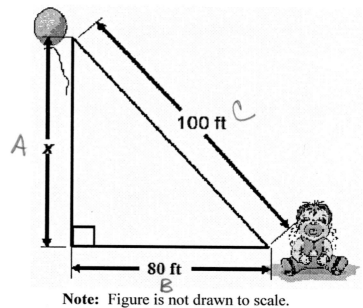

Note: Figure is not drawn to scale.

What we need to do here is figure the height of the right triangle. We can do this by using Pythagorean's Theorem, which is:

$$a^2 + b^2 = c^2$$

a and b are the two shortest sides. c is the longest side, the side across from the 90 degree corner. This longest side is known as the hypotenuse.

We can substitute the 80 in for either a or b. We have to substitute the 100 in for c since it is the longest side.

$$80^2 + b^2 = 100^2$$

Next, we square the 80 and the 100. If you do not have a calculator, your reference page has a table with the squares of the numbers 1 – 10.

$$6400 + b^2 = 10,000$$

Now, we subtract 6400 from both sides to get the b^2 by itself.

$$
\begin{array}{r}
6400 + b^2 = 10,000 \\
-6400 \qquad\quad -6400 \\
\hline
b^2 = 3600
\end{array}
$$

Our last step, to get the b by itself, is to take the square root of both sides. (The opposite of squaring a number is to take the square root.)

$$\sqrt{b^2} = \sqrt{3600} \;\rightarrow\; b = 60$$

The correct answer is G.

Another way to find the answer to this one is to substitute the answers in for b and see if the left side of the equation equals the right side.

F. $30 \rightarrow 80^2 + 30^2 = 100^2 \rightarrow 6400 + 900 = 10,000 \rightarrow 7300 \neq 10,000$

Using order of operations on the left side we get 7300. 7300 does not equal 10,000 so F is not the correct answer. Now, we will try G.

G. $60 \rightarrow 80^2 + 60^2 = 100^2 \rightarrow 6400 + 3600 + 10,000 \rightarrow 10,000 = 10,000$

On this one we get 10,000 equals 10,000. Both sides are equal so that is our correct answer.

53. What is the slope of the line on the graph?

A. $\dfrac{2}{3}$

B. $-\dfrac{2}{3}$

C. $\dfrac{3}{2}$

D. $-\dfrac{3}{2}$

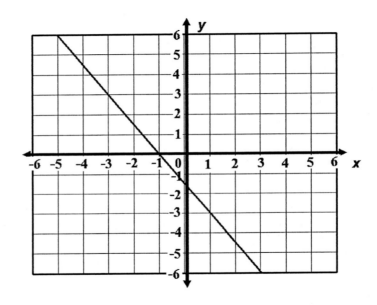

The first thing we want to look at is the direction the line is going. If the line is going up, from left to right, the slope is positive. If the line is going down, from left to right, the slope is negative.

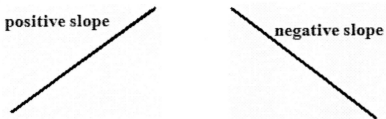

We can eliminate answers A and C because they are both positive.

119

There are two ways that we can find the slope of this line. One way is to use the slope formula that is on the reference page.

$$Slope = \frac{y_2 - y_1}{x_2 - x_1}$$

We just need to pick two points on the line. It can be any two points on the line.

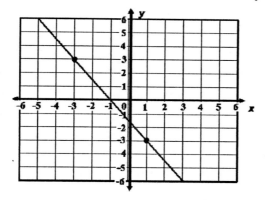

I have chosen the points (-3, 3) and (1, -3). I will use the point (-3, 3) as point 1, and (1, -3) as point 2. Using the formula, we will substitute the numbers from the points into the formula. (Remember, the first number in a point is the x and the second number is the y.)

$$Slope = \frac{y_2 - y_1}{x_2 - x_1} \rightarrow \frac{-3 - 3}{1 - (-3)} = \frac{-6}{4} = \frac{-3}{2}$$

We get 4 in the denominator (bottom number) because when we have two negative signs together, it is the same as addition.

The correct answer is D.

Another way of figuring the slope is to make a right triangle, using the line on the graph as the hypotenuse.

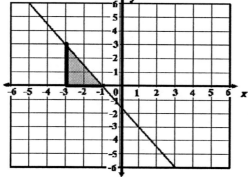

Don't forget, another way of looking at slope is rise over run. The distance we go on the vertical line (the line that goes up) is our top number. The distance we go on the horizontal line (the line that goes sideways) is our bottom number. We go a total of 3 places on the vertical line, so the

120

top number is 3. We go two places on the horizontal line, so the bottom number is 2. Since the line is going down hill, the sign is negative.

So our answer is $-\dfrac{3}{2}$. Again the correct answer is D.

54. Which of these graphs represents $y = 2x - 3$?

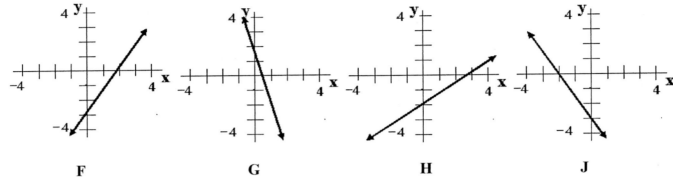

F	**G**	**H**	**J**

This one is easy because it is in slope intercept form which is y = mx + b. The m tells us our slope and the b tells us where it crosses the y axis. This line crosses the y axis at – 4. On my graph, I will count down 4 places from the center and make a point.

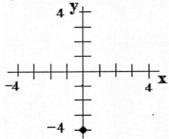

Now, using the number in front of the x, which is our slope, we can figure out where our second point belongs. When looking at the slope, think rise over run. The top number tells us how far up or down to go (if the slope is positive, go up, if negative, go down), and the bottom number tells us how far to go to the right (we always go to the right). If the slope is not a fraction, we can put the number over one to make into a fraction.

$\dfrac{2}{1}$ The slope is positive so we will go up 2 places from our point and to the right 1 place.

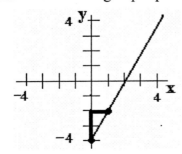

We draw a line through the two points to complete our graph. This graph looks most like choice A. **The correct answer is F.**

121

55. A contractor is planning to run an underground line diagonally across a field. The shortest distance is a straight line. How much line will he need to go across the field?

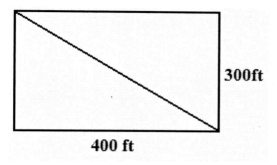

A. 200 ft

B. 300 ft

C. 400 ft

D. 500 ft

The diagonals (opposite corners) of a rectangle are the same no matter which two opposite corners you use. Some students get confused on this problem because they don't see that they can use Pythagorean's Theorem (it is on the reference page). What we really have here is a right triangle problem.

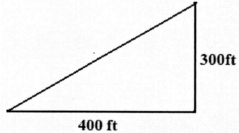

We use the formula $a^2+b^2 = c^2$. The two shortest sides are your a and b. It doesn't matter which one you use as a or which one as b. The hypotenuse (the longest side) is always c. Now, we will substitute the 400 in for a and 300 in for b and use order of operations. We will square the number first, then add them.

$$a^2+b^2 = c^2 \rightarrow 400^2+300^2 = c^2 \rightarrow 160,000 + 90,000 = c^2 \rightarrow 250,000 = c^2$$

To get our final answer we need to take the square root of each side. We do this because the c is squared, and we want to solve for c.

$$\sqrt{250,000} = \sqrt{c^2} \rightarrow 500 = c$$

The correct answer is D.

56. The following graph represents the equation y = 2x + 2.

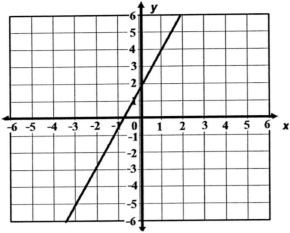

If the constant (2) changes from positive to negative, what will the graph look like?

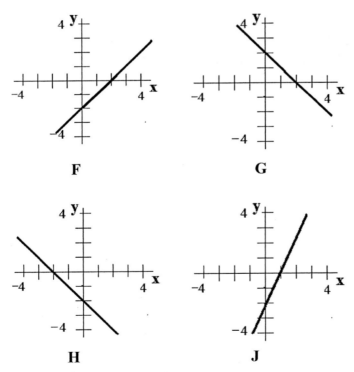

This problem actually looks more complicated than it really is. We are just changing the last number in the equation. Remember, when the equation is in the form of y = mx + b, the last number is where the line crosses the y-axis. So, if we change the + 2 to – 2 that means the line will cross the y-axis at – 2. The slope stays the same. The line simply moves down to where it crosses at – 2.

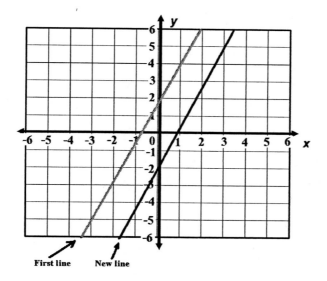

First line   New line

The new line is in the same position as answer J.

The correct answer is J.

57. A gardener is mixing bags of potting mix. He makes two kinds. In the first kind he uses 4 pounds of compost and 3 pounds of fine sand. In the second kind he uses 6 pounds of compost and 8 pounds of fine sand. The gardener has 140 pounds of compost and 140 pounds of fine sand. He wants to make the maximum possible number of batches of potting mix and use as much of the compost and fine sand as he can. How many batches of each kind of mix can he make?

  A. 0 batches of the first and 30
     batches of the second

  B. 20 batches of the first and 10
     batches of the second

  C. 15 batches of the first and 15
     batches of the second

  D. 15 batches of the first and 20
     batches of the second

The easiest way of solving this problem is to try each answer and see which one is best.

A. 0 batches of the first and 30 batches of the second

The 2nd mix uses 6 pounds of compost and 8 pounds of fine sand. We will multiply 6 times 30 to find out how much compost he would use, and 8 times 30 to find out how much fine sand he would use.

6 x 30 = 180 pounds of compost. He only has 140 pounds so that can't be right.

B. 20 batches of the first and 10 batches of the second

In the first kind he uses 4 pounds of compost and 3 pounds of fine sand. In the second kind he uses 6 pounds of compost and 8 pounds of fine sand. We need to see how much compost and fine sand he will use making each type of mix.

Mix one: 4 x 20 = 80 pounds of compost
          3 x 20 = 60 pounds of fine sand

Mix two: 6 x 10 = 60 pounds of compost
          8 x 10 = 80 pounds of fine sand

Now, we add the amount of compost for the first and second type of mix to get the total amount used. 80 + 60 = 140 pounds.
Then, we add the amount of fine sand used for the first and second types of mix to get the total amount used. 60 + 80 = 140

The gardener wanted to use as much as possible of the fine sand and compost. Since he is using all of both, it appears that this is the correct answer. I am going to check answer D. The reason I want to check D is because that one makes 35 batches of mix. Remember, he wanted to make as many batches as possible and use as much mix as possible.

D. 15 batches of the first and 20 batches of the second

Mix one: 4 x 15 = 60 pounds of compost
          3 x 15 = 45 pounds of fine sand

Mix two: 6 x 20 = 120 pounds of compost
          8 x 20 = 160 pounds of fine sand

This can't be the correct answer because the second mix needs 160 pounds of fine sand and the gardener only has 140 pounds.

The correct answer is B.

58. What is the reciprocal of 1?

    F. $\dfrac{1}{2}$

    G. -1

    H. $\dfrac{2}{1}$

    J. 1

To find the reciprocal, we flip the fraction over. If the number is not a fraction, we make it a fraction by putting the number over 1. So, to make 1 into a fraction we will put 1 under it and we get:
$$\frac{1}{1}$$
Now, we flip the fraction over to get the reciprocal. When we flip it over, we end up with:
$$\frac{1}{1}$$
1 over 1 equals 1. So, the reciprocal of 1 is 1.

The correct answer is J.

59. Multiply: $(2x - 3)(2x - 5)$

    A. $4x + 15$

    B. $4x^2 - 15$

    C. $4x^2 - 5x - 15$

    D. $4x^2 - 16x + 15$

This one is like question 41. We need to multiply the 2x in the first set of parenthesis by both of the terms in the second set of parenthesis. So, we will multiply the 2x times 2x and 2x times – 5.

$$(2x - 3)(2x - 5)$$

$$2x \cdot 2x = 4x^2 \qquad 2x \cdot (-5) = -10x$$

Next, we multiply the – 3 times both the terms in the second set of parenthesis. So, we will multiply – 3 times 2x and – 3 times – 5.

$$(2x - 3)(2x - 5)$$

$$-3 \cdot 2x = -6x \qquad -3 \cdot (-5) = 15$$

Remember, when multiplying, if the signs in front of the numbers are the same, the answer is positive. If the signs are different, the answer is negative. That is why our first answer is negative, and our second answer is positive.

Putting it all together we have:

$$(2x - 3)(2x - 5)$$

$$4x^2 - 10x - 6x + 15$$

As we do the multiplication, we write down our answers from left to right as we go. Our last step is to combine our two middle terms. We have $-10x - 6x = -16x$. Our middle term is now – 16x. So, our final answer is:

$$4x^2 - 16x + 15$$

Many of you were probably taught to **Foil** the two terms. Foil stands for **first, outer, inner,** and **last.** Here's how it works.

First	**Outer**	**Inner**	**Last**
Multiply the first numbers together	Multiply the two outer numbers	Multiply the two inside numbers	Multiply the last two numbers
$(2x - 3)(2x - 5)$	$(2x - 3)(2x - 5)$	$(2x - 3)(2x - 5)$	$(2x - 3)(2x - 5)$
$(2x)(2x) = 4x^2$	$(2x)(-5) = -10x$	$(-3)(2x) = -16x$	$(-3)(-5) = -15$

Next, we list our answers from left to right. Be sure if there is a negative sign in front of the number that you bring it with the number in the problem.

$$4x^2 - 10x - 6x + 15$$

Our last step is to combine our two middle terms. We have $-10x - 6x = -16x$. Our middle term is now $-16x$. So, our final answer is:

$$4x^2 - 16x + 15$$

The correct answer is D.

60. Workers are pouring a sidewalk around a rectangular garden. The outer perimeter of the sidewalk needs to be at least 116 feet. If the length of the sidewalk must be 9 feet longer than the width, what is the least possible integer value of the length of the sidewalk?

    F. 34 ft

    G. 27 ft

    H. 25 ft

    J. 23 ft

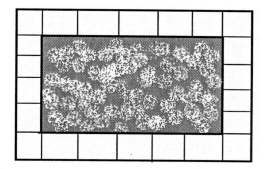

On the reference page we can find the formula for the parameter of a rectangle.

Parameter $= 2l + 2w$

Now, we have to read this one carefully and use the information that is given to us. We do not know how long or how wide the rectangle is. We are told that the length has to be 9 feet wider than the width.

We can use w to represent the width. Since we are told the length has to be 9 ft longer, our length is $w + 9$. So our width is w and our length is $w + 9$. Since we have our length and width, we can use the formula $2l + 2w$. We need to remember to put the $w + 9$ inside parenthesis. (Anytime we have two or more terms together we put them in parenthesis to keep them together.) We will also put in 116 for the parameter. So, we have:

$$116 = 2(w + 9) + 2w + 9$$

Now, we get rid of the parenthesis by multiplying 2 times the w and 9 inside the parenthesis.

$$116 = 2(w + 9) + 2w$$

2 times w equals 2w, and 2 times 9 equals 18. So, now we have:

$$116 = 2w + 18 + 2w$$

Our next step is to combine our like terms. We will combine our w's and our two numbers. 2w + 2w = 4w, and we have + 18. So, now we have:

$$116 = 4w + 18$$

Next, we need to get the w by itself. Remember, we move whatever is farthest away from the w first. The 18 is farther from the w than the 4. The 116 is already on the other side of the equal sign, so it does not need to be moved. We will move the 18 by subtracting 18 from each side of the equal sign.

$$\begin{array}{rcl} 116 &=& 4w + 18 \\ -18 & & \quad -18 \\ \hline 98 &=& 4w \end{array}$$

Our last step is to divide both sides by 4.

$$\frac{98}{4} = \frac{4w}{4}$$

The 4's on the right cancel out. 89 divided by four is 24.5. We have to round this number up to the next highest number. So, the width is 25. Some students stop there, and make that their answer. The question is asking for the width. The length is 9 feet longer than the width. 25 + 9 = 34.

The correct answer is F.

We could use the answers that are given to us. We will start with the lowest answer and work our way up. It helps to have a picture to work with. So, let's draw a rectangle.

23

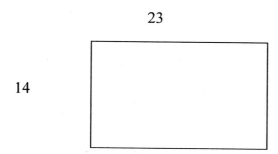

14

We will use 23 feet for the length, because that is the lowest number, and the answer gives us the length not the width. We know that the length is 9 feet longer than the width, so we need to subtract 9 from 23 and get 14 for the width.

Now, we use the formula $2l + 2w$. We substitute 23 in for the l and 14 in for the w.

$$2(23) + 2(14) = 46 + 28 = 74$$

Using order of operations, we get 74 as the parameter. The parameter has to be at least 116 feet.

Next, let's try 25 for the length. Using 25 for the length we get 16 for the width, after subtracting 9.

$$2(25) + 2(16) = 50 + 32 = 82$$

Again, we need a parameter of at least 116, so that is not our answer.

Now, we are going to try 27 for our length and 18 for the width.

$$2(27) + 2(18) = 54 + 36 = 90$$

In this case it would have been quicker to start at the highest number. But we have to be careful because we are looking for the smallest length that will give us at least 116 feet. So, now we are going to use 32 feet for the length and 23 feet for the width.

$$2(34) + 2(25) = 68 + 50 = 118$$

Even though this one goes a couple of numbers above 116, it is the smallest whole number that will give us at least a 116 foot parameter.

The correct answer is F.

61. How many different ways can 7 books be arranged on a shelf?

    A. 49

    B. 70

    C. 7!

    D. infinitely many

Many times in Algebra I, teachers do not get to the point in the curriculum where they teach about factorials. Or, they teach it a different way and do not use the "!" sign in their definitions. If we have a number with the exclamation sign next to it, we multiply the number times all the numbers behind it. Example:

$$4! = 4 \times 3 \times 2 \times 1 = 24$$

The way that many students are taught to solve this problem is to imagine 7 places on a shelf and looking at how many choices there are for each slot.

— — — — — — —

We have our 7 places where we can put our books. Now, we ask ourselves, how many choices do we have for the first position? Well, we could put any of the 7 books there, so our answer for the first position is 7.

   7 __ __ __ __ __ __

Next, we look at how many books we have left to choose to put in the next spot. After we put one book in the first place, we have a total of 6 books left. So, the next one would be 6.

   7 6 __ __ __ __ __

We continue in the same manner until we have filled all the places. We subtract 1 number each time. When we fill all the slots, it will look like this:

   7 6 5 4 3 2 1

To get the total number of different ways the books could be arranged on the shelf, we multiply all the numbers together.

$$7 \times 6 \times 5 \times 4 \times 3 \times 2 \times 1 = 5040$$

5040 is not one of the choices, but we know that 7! is the same as $7 \times 6 \times 5 \times 4 \times 3 \times 2 \times 1$. **Therefore, the correct answer is C.**

62. Which of the following figures shows an area representation of $(2x + 3)$ multiplied by $(x + 1)$?

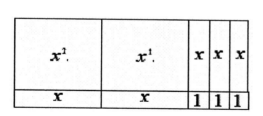

F

G

H

J

Be prepared, there will most likely be an algebra tile question. I know of very few teachers who still use algebra tiles in the classroom. We can simply perform the multiplication, get our answer, and then match it with the algebra tile answer.

This one is like questions 41 and 59. We need to multiply the 2x in the first set of parenthesis by both of the terms in the second set of parenthesis. So, we will multiply the 2x times x and 2x times 1.

$$(2x + 3)(x + 1)$$

$$2x \cdot x = 2x^2 \qquad 2x \cdot 1 = 2x$$

Next, we multiply the + 3 times both the terms in the second set of parenthesis. So, we will multiply + 3 times x and − 3 times 1.

$$(2x + 3)(x + 1)$$

$$3 \cdot x = 3x \qquad 3 \cdot 1 = 3$$

Remember, when multiplying, if the signs in front of the numbers are the same, the answer is positive. If the signs are different, the answer is negative.

Putting it all together we have:

$$(2x + 3)(x + 1)$$

$$2x^2 + 2x + 3x + 3$$

As we do the multiplication, we write down our answers from left to right as we go.
Our last step is to combine our two middle terms. We have $+2x + 3x$. Our middle term is now 5x. So, our final answer is:

$$2x^2 + 5x + 3$$

Now, we need to match our answer with the algebra tile answers given. Looking at the first term in our answer, we have $2x^2$. This means we will have 2 of the squares with the x^2 in them. All of the answers have two of the squares so we cannot eliminate any of our answers yet.

Next, we will look at our middle term, 5x. This means that we will have 5 of the rectangles with an x in them. F is the only answer with 5 rectangles.

Let's go ahead and check our last term, 3. This means that there will be 3 small squares with a 1 in them. Answer F also has the three squares with a 1 in each of them.

The correct answer is F.

A few more examples.

1. Andre is packing glass jars that will be used to distribute jelly. The following table shows the number of jars he had packed throughout the day.

Time	Number of Jars
10:00 am	50
12:00 pm	100
2:00 pm	150
4:00 pm	200

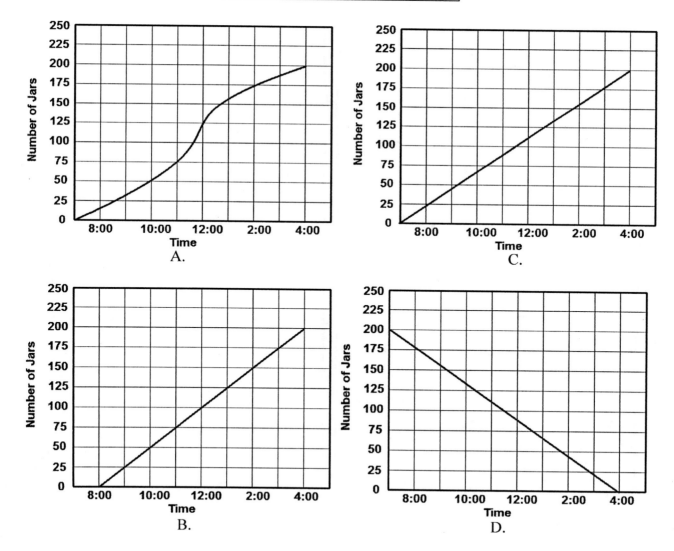

A.

C.

B.

D.

134

2. What is the range of the graph?

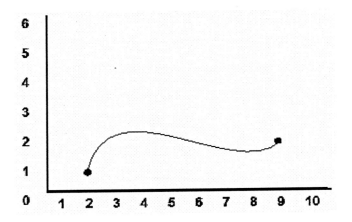

F. $1 \le R \le 2$

G. $2 \le R \le 9$

H. $2 \ge R \le 9$

J. $1 \ge R \le 2$

3. What is the domain of the graph in question 2?

A. $1 \le R \le 2$

B. $2 \le R \le 9$

C. $2 \ge R \le 9$

D. $1 \ge R \le 2$

4. Simplify: $4x(2 - x^2)$

F. $6x - 4x^3$

G. $2x$

H. $8x - 4x^3$

J. $4x^2$

5. Which of the following is "a number n divided by seven equals three more than a number n."

A. $\dfrac{7}{n} + 3 = n$

C. $\dfrac{n}{7} = n + 3$

B. $\dfrac{n}{7} = n - 3$

D. $\dfrac{7}{n} = n + 3$

6. What is the volume of a packing crate with the dimensions: 5' by 2 ½ ' by 2 ½ '?

F. $10\,ft^2$

G. $10\,ft^3$

H. $25\,ft^3$

J. $31.25\,ft^3$

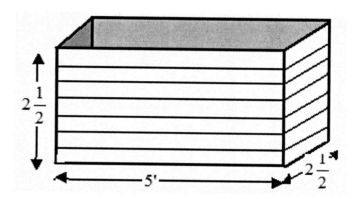

Answers

Test 1

1. C
2. J
3. C
4. F
5. A
6. J
7. D
8. F
9. B
10. J
11. D
12. G
13. D
14. F
15. B
16. F
17. B
18. H
19. D
20. H
21. D
22. H
23. D
24. J
25. A
26. G
27. A
28. J
29. B
30. G
31. D
32. H
33. B
34. G
35. D
36. F
37. C
38. J
39. D
40. H
41. A
42. F
43. D
44. G
45. A
46. J
47. C
48. H
49. A
50. G
51. A
52. H
53. D
54. G
55. B
56. F
57. B
58. J
59. B
60. J
61. C
62. G

Test 2

1. C
2. G
3. A
4. J
5. D
6. F
7. C
8. G
9. B
10. H
11. C
12. F
13. A
14. G
15. C
16. H
17. C
18. F
19. C
20. G
21. B
22. G
23. D
24. H
25. A
26. H
27. B
28. G
29. C
30. H
31. B
32. H
33. D
34. F
35. B
36. J
37. B
38. G
39. C
40. J
41. B
42. F
43. B
44. G
45. D
46. J
47. A
48. H
49. A
50. G
51. B
52. F
53. B
54. G
55. D
56. F
57. B
58. G
59. C
60. J
61. A
62. F

Test 3

1. B
2. H
3. C
4. J
5. A
6. H
7. D
8. G
9. B
10. H
11. D
12. G
13. B
14. F
15. B
16. H
17. A
18. F
19. D
20. J
21. C
22. F
23. D
24. J
25. B
26. H
27. B
28. H
29. C
30. G
31. D
32. H
33. B
34. G
35. C
36. J
37. A
38. H
39. D
40. G
41. A
42. H
43. B
44. H
45. C
46. F
47. B
48. H
49. D
50. G
51. B
52. G
53. D
54. F
55. D
56. J
57. B
58. J
59. D
60. F
61. C
62. F

Extra Problems

1. B
2. F
3. G
4. H
5. C
6. J

Printed in the United States
31639LVS00004B/19

9 780976 638001